THE PIPE FITTERS BLUE BOOK

REVISED

By W. V. GRAVES

Formerly Pipe Fitter Instructor
LEE COLLEGE
BAYTOWN, TEXAS

ORDER FROM:

GRAVES PUBLISHING CO.
P. O. BOX 57924
WEBSTER, TEXAS 77598

(281) 480-2405

Publisher of
The Pipe Fitters Blue Book
And The Pipe Fabricators Blue Book

ISBN 0-9708321-0-9

PREFACE AND INSTRUCTIONS

The author has included in this book the information and charts most often needed on piping jobs.

The explanations and methods used have been made as simple as possible so that you should have little difficulty in understanding them.

Many of the cuts for fabrication in this book are based on the inside diameter of the riser or branch to be fitted onto the outside diameter of the header and should be cut radially with the torch cutting tip pointed toward the center of the pipe at all times.

After cutting, the risers may then be placed in position on the header for marking the header cut line.

Miter cuts should be cut with the cutting tip pointed into the line as though you were using a saw cut.

Pipe may be marked off in quarters, eighths, or sixteenths by using the table in this book, or by folding a piece of paper that has been fitted around the circumference of the pipe so that the ends of this paper just meet.

The wraparound should be carefully fitted onto the pipe and kept square so that you will have a true reference line.

Note that the fabrication charts are calculated for the use of standard weight and extra strong wall thickness pipe and are accurate for these wall thicknesses only.

TABLE OF CONTENTS

COMMON PIPING ANGLES AND THEIR SOLUTIONS

NOTE THAT ALL NINE OF THE FOLLOWING DRAWINGS SHOW A RIGHT TRIANGLE. The pipe fitter usually calls the lengths of their sides (SET), (RUN) and (TRAVEL).

These terms may be used to find the angles as well as the lengths of the sides, by referring to pages 9 and 10 of this book.

DRAWING #1 Shows a 30° offset. The level run of pipe intersects the (TRAVEL) at a 30° angle. If the length of the (SET) is known, the lengths of the (RUN) and (TRAVEL) may be found by referring to page 10 under (ANGLE KNOWN) in the 30° column. These formulas may be used for any angle not shown in this table by use of the trigonometry tables in the back of this book.

DRAWING #2 Shows the same triangle as before however the pipe now is vertical and intersects the 60°angle. To find the lengths of the (RUN) and (TRAVEL) refer to page 10 under (ANGLE KNOWN) in the 60° column. Note that when the (SET) side is longer than the (RUN) side the angle will always be more than a 45° angle.

DRAWING #3 Shows a vessel with a nozzle that is 30° over from a reference line. If the dimension from the face of nozzle to the centerline of the vessel is known you would add the laying length of a 30° weldell plus a welding neck flange. See drawing #8 for method of calculating the laying length of a 30° weldell. These dimensions added together gives you the length of the (TRAVEL) side. To find the (SET) and (RUN) sides refer to page 10 under (ANGLE KNOWN).

DRAWING #4 Shown are two 60° offsets that are staggered so that equal spacing will be maintained at all centerlines of of the pipe. Note that there are two 30° triangles shown. FORMULA FOR STAGGERING OFFSETS = CENTER TO CENTER DISTANCE OF PIPE x TANGENT OF ½ THE DEGREES OF TURN OF OFFSET. The figure for 60° is .577; for 45° is .414; for 30° is .268.

COMMON PIPING ANGLES AND THEIR SOLUTIONS

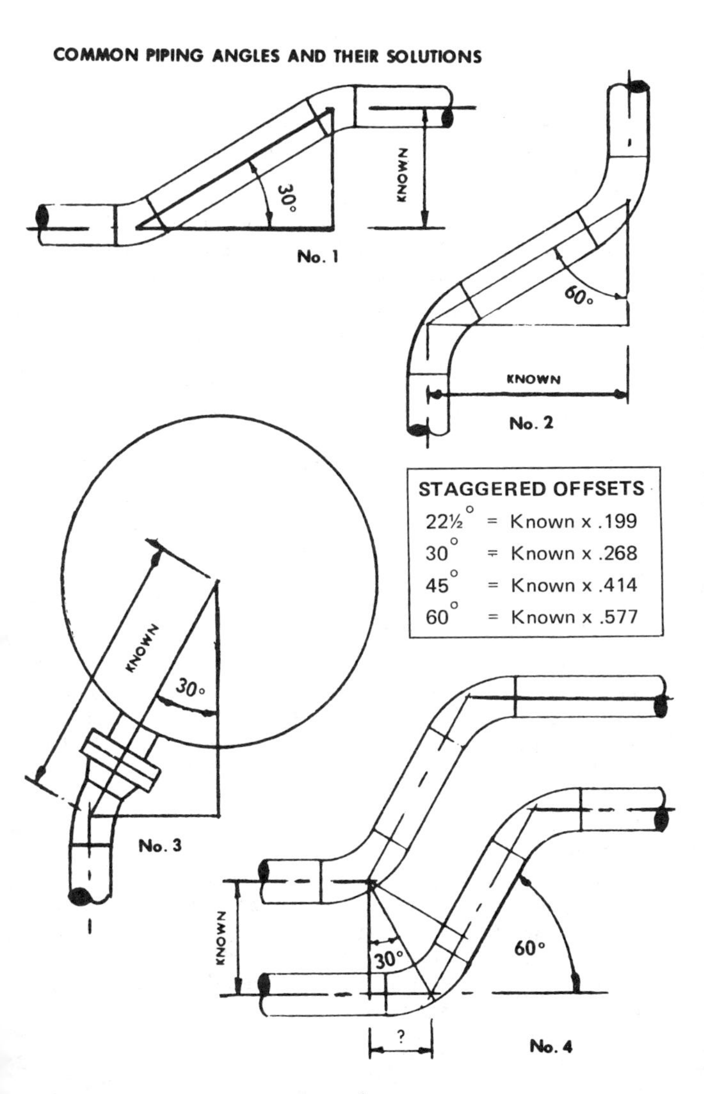

STAGGERED OFFSETS		
22½°	=	Known x .199
30°	=	Known x .268
45°	=	Known x .414
60°	=	Known x .577

DRAWING #5 Shows the right triangle formed for a miter cut on pipe. The angle of cut is ½ of the degrees of turn. FORMULA FOR MITER CUT ON PIPE = O.D. OF PIPE x TANGENT OF ANGLE OF CUT. Usually a single wraparound mark is drawn on pipe and ½ of the above dimension is marked off on each side of this line. Refer to pages 13 through 19 for examples of layout and calculated dimension tables.

DRAWING #6 Shows a single piece of angle iron that is cut and then bent to form a one piece turn. On this type mark off a centerline and layout a cutback on each side of this line as shown. Type shown has a turn of 135° and requires two 67½° cutbacks. FORMULA FOR CUTBACK = WIDTH MINUS ONE THICKNESS x TANGENT OF ANGLE OF CUT. Refer to angle iron brackets and tables of calculated dimensions for additional information.

DRAWING #7 Shows a piece of angle iron that is cut across the total width. Two pieces are required for a turn. FORMULA FOR CUTBACK = WIDTH x TANGENT OF ANGLE OF CUT. The cutback for other types of steel may be found with these formulas or they can be marked off with a protractor.

DRAWING #8 Shows a pipe turn of 60° and the two right triangles formed to determine the laying length or end to center. As the radius is generally known for weldells and bends, the end to center dimension can be calculated. FORMULA = RADIUS x TANGENT OF ½ THE DEGREES OF FITTING AND OR BEND. See table of calculated dimensions of long radius weldells in this book.

DRAWING #9 Shows a 90° weldell rolled over 45° with a 45° weldell added to run pipe level. The centers of these fittings added together form the (TRAVEL) side of a right triangle. The (SET) and (RUN) sides may be calculated by referring to page 10 under (ANGLE KNOWN). All types of fittings or combinations of fittings that are rolled over on angles may be solved by this procedure.

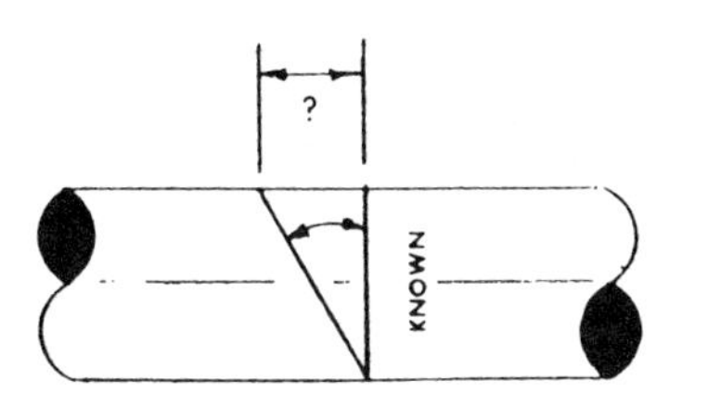

No. 5

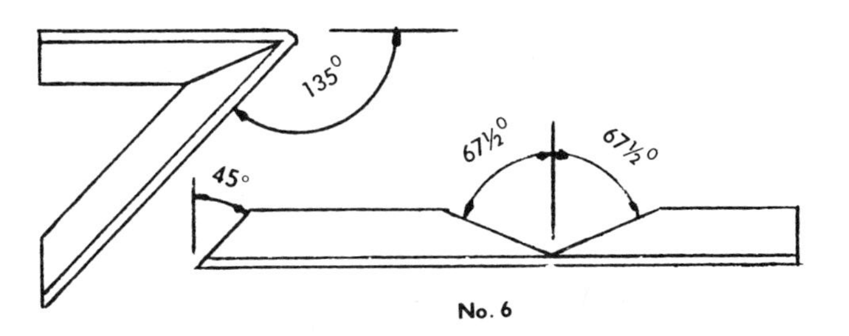

No. 6

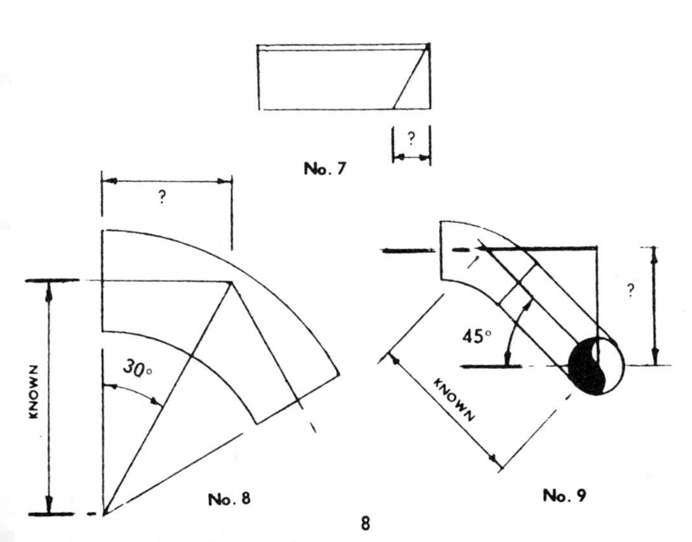

No. 7

No. 8

No. 9

SOLUTION OF ANGLES BY THE USE OF TERMS FAMILIAR TO THE PIPE FITTER

Piping angles and the lengths of their sides may easily be found by using the following methods. These are based on solving the angles and the lengths of sides of a right triangle. A right triangle has three angles which add to 180°. As one angle is always 90° the sum of the other 2 angles always add to 90°.

RULE FOR FINDING AN UNKNOWN ANGLE OR LENGTH OF AN UNKNOWN SIDE

To find an unknown angle you must know the lengths of any two sides, such as the (SET) and (RUN).

To find the length of an unknown side you must know the angle and the length of one side.

HOW TO USE THE TRIGONOMETRY TABLES IN THE BACK OF THIS BOOK

For angles of 45° or less read the angles, constants, and minutes of a degree from the top of the page down.

For angles of 45° or more read the angles, constants, and minutes of a degree from the bottom of the page up.

HOW TO FIND AN UNKNOWN ANGLE

EXAMPLE: Refer to the piping offset on page 6. If the (SET) length is 10" and the (TRAVEL) length is 20" what is the angle of this offset? The table TO FIND ANGLE is used and it shows that the (SET) divided by the (TRAVEL) = the SINE of this angle. 10 divided by 20 = .500 and by looking for this figure in the SINE column of the trigonometry tables it is found to be a 30° angle. The table also shows that the (TRAVEL) divided by the (SET) = the COSECANT of this angle.

HOW TO FIND THE LENGTH OF AN UNKNOWN SIDE

The offset above has an angle of 30°, the (SET) is 10" and the (TRAVEL) is 20". What is the length of the (RUN) side?

EXAMPLE: Refer to the table ANGLE KNOWN and it shows that the (RUN) = the (TRAVEL) × the (COSINE) or 20 × .866=17.32 or 17 5/16". Also the table shows that the (RUN) = the (SET) × (COTANGENT).

NOTE: If the (SET) and (RUN) lengths are the same, the angle is 45°.

If the (SET) is less than the (RUN), the angle is less than 45°.

SHOW THE ANGLE ON YOUR DRAWING BETWEEN THE (TRAVEL) AND (RUN).

(SET) = side opposite, (RUN) = side adjacent, (TRAVEL) = hypotenuse.

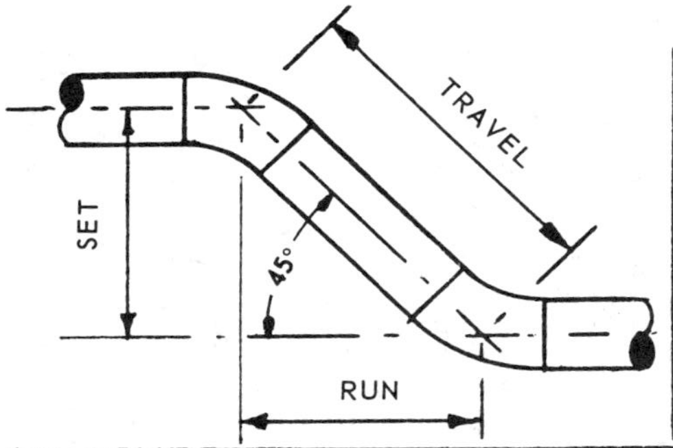

HOW TO FIND THE ANGLE WHEN THE LENGTHS OF 2 SIDES ARE KNOWN.

SET DIVIDED BY TRAVEL = SINE
RUN DIVIDED BY TRAVEL = COSINE
SET DIVIDED BY RUN = TANGENT
RUN DIVIDED BY SET = COTANGENT
TRAVEL DIVIDED BY RUN = SECANT
TRAVEL DIVIDED BY SET = COSECANT

TO FIND LENGTHS OF SIDES WHEN THE ANGLE IS KNOWN	ANGLE OF OFFSET							
	60°	45°	30°	22½°	15°	11¼°	9°	7½°
SET = TRAVEL × SINE	.866	.707	.500	.383	.259	.195	.156	.130
RUN = TRAVEL × COSINE	.500	.707	.866	.924	.966	.981	.988	.991
SET = RUN × TANGENT	1.732	1.000	.577	.414	.268	.199	.158	.132
RUN = SET × COTANGENT	.577	1.000	1.732	2.414	3.732	5.027	6.314	7.596
TRAVEL = RUN × SECANT	2.000	1.414	1.155	1.082	1.035	1.020	1.012	1.008
TRAVEL = SET × COSECANT	1.155	1.414	2.000	2.613	3.864	5.126	6.392	7.661

SOLVING ROLLING OFFSETS

A rolling offset is nothing more than a plain offset rolled over so as to hold 2 dimensions as shown in drawing (a).

To figure a rolling offset you must find the distance it will take to roll over straight piping to hold these 2 dimensions. This is shown in drawing (a) as piping with 90° turns. When this distance is found the offset is figured the same as a simple offset. The easy way to find this distance is to measure across a steel square at 10" and 21¼" as shown in drawing at (a). The figure of 23½" that is obtained is the true amount of (SET) and is shown again in true view at drawing (b). To find the amount of (TRAVEL) needed for any angle multiply this (SET) of 23½" x the COSECANT of the angle you are using. In the drawing at (b) a 45° angle is shown. For a 45° angle multiply 23½" x 1.414 = 33¼". For a 30° angle multiply 23½" x2.000 = 47".

This distance of 23½", may also be calculated by the use of square root; or may be calculated if the angle formed by the 10" (SET) and 21¼" (RUN) is found. Refer to the table (TO FIND ANGLE) 21¼" divided by 10" = 2.125 This figure is the CONTANGENT of 25° 12'. Now that the angle is known refer to the table (ANGLE KNOWN) and note that the TRAVEL = SET X COSECANT or 10" x 2.3486 = 23½". This is the same figure that was obtained on the steel square. This method can be used if the distances are greater than the length of a steel square.

In order to figure a rolling offset that will also hold to a given dimension as at (c) in drawing (a). This can be done simply. Assume that dimension (c) is to be 18" refer to view (b) and note that 23½" is the (SET). The 18" side is the (RUN). Use the lengths of these 2 sides to figure the angle. Refer to the table (TO FIND ANGLE). It shows that the SET DIVIDED BY RUN = TANGENT or 23½" divided by 18" = 1.30555. The trig tables show this to be the Tangent of 52° 33'. For practical purposes call it 52°30' and calculate the lengths of the (RUN) and (TRAVEL) sides for this degree. See table (ANGLE KNOWN). The The (RUN) side equals 23½" x .7673 = 18". The(TRAVEL) side equals 23½" x 1.2605 = 29⅝". When the (SET) side is greater than the (RUN) side the angle will be more than 45° angle and the angles in the trig tables are read from the bottom of the page up.

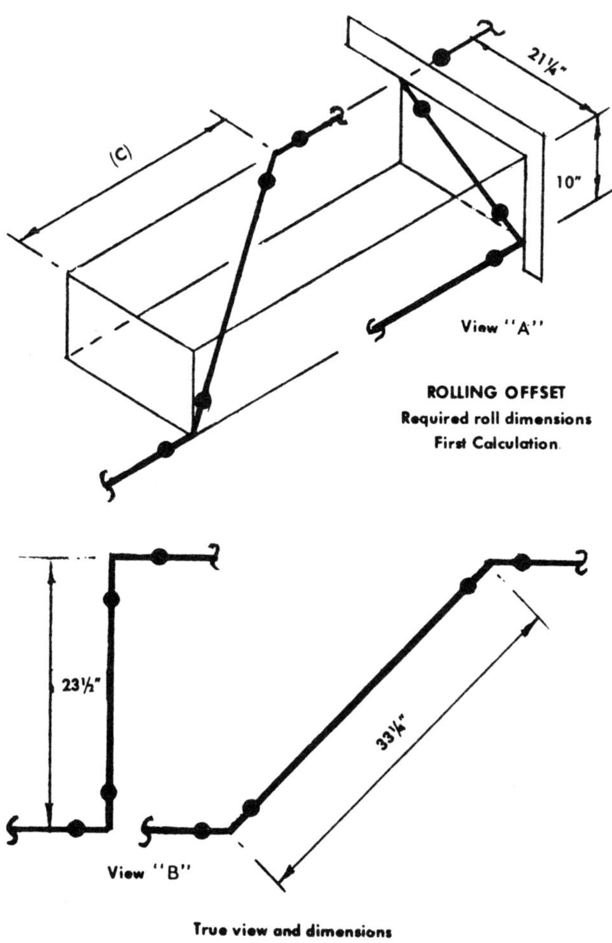

ROLLING OFFSET
Required roll dimensions
First Calculation.

True view and dimensions
of Rolling Offset for above
Second Calculation

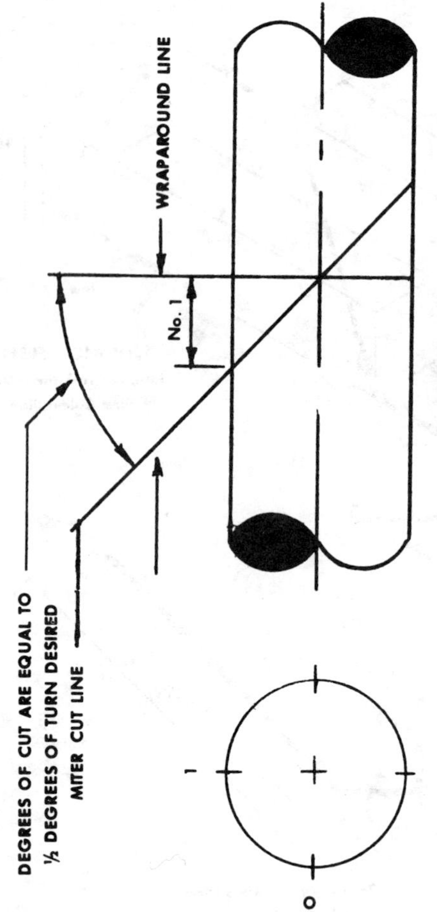

MITER CUTS FOR 1½" THROUGH 3" WITH PIPE MARKED IN QUARTERS.
LINE No. 1 DIMENSION EQUALS TANGENT OF CUT X O.D. OF
PIPE DIVIDED BY 2

1½" THROUGH 3" MITER CUTS PIPE QUARTERED

7½° CUT FOR 15° TURN	
SIZE	NO. 1
1½	1/8
2	1/8
2½	3/16
3	3/16

9° CUT FOR 18° TURN	
SIZE	NO. 1
1½	1/8
2	3/16
2½	1/4
3	1/4

11¼° CUT FOR 22½° TURN	
SIZE	NO. 1
1½	3/16
2	1/4
2½	1/4
3	5/16

15° CUT FOR 30° TURN	
SIZE	NO. 1
1½	1/4
2	5/16
2½	3/8
3	7/16

22½° CUT FOR 45° TURN	
SIZE	NO. 1
1½	3/8
2	1/2
2½	9/16
3	3/4

30° CUT FOR 60° TURN	
SIZE	NO. 1
1½	1/2
2	11/16
2½	13/16
3	1

45° CUT FOR 90° TURN	
SIZE	NO. 1
1½	15/16
2	1 3/16
2½	1 7/16
3	1 3/4

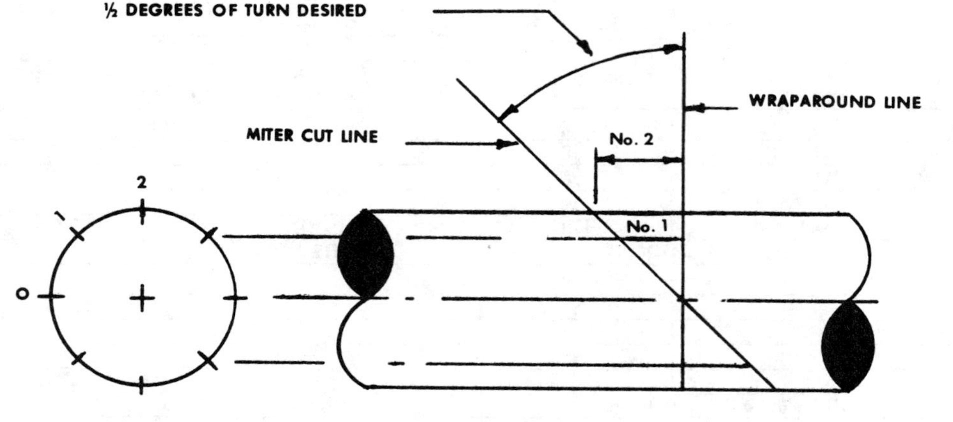

MITER CUTS FOR 4" THROUGH 10" WITH PIPE MARKED IN EIGHTHS
LINE No. 2 DIMENSION EQUALS TANGENT OF CUT X O.D. OF PIPE
DIVIDED BY 2
LINE No. 1 DIMENSION EQUALS DIMENSION No. 2 X .7071

4" THROUGH 10" MITER CUTS PIPE IN EIGHTHS

7½° CUT FOR 15° TURN		
SIZE	NO. 1	NO. 2
4	3/16	1/4
6	5/16	7/16
8	3/8	9/16
10	1/2	11/16

9° CUT FOR 18° TURN		
SIZE	NO. 1	NO. 2
4	1/4	3/8
6	5/16	1/2
8	1/2	11/16
10	5/8	7/8

11¼° CUT FOR 22½° TURN		
SIZE	NO. 1	NO. 2
4	5/16	7/16
6	7/16	5/8
8	5/8	7/8
10	3/4	1 1/16

15° CUT FOR 30° TURN		
SIZE	NO. 1	NO. 2
4	3/8	9/16
6	5/8	7/8
8	13/16	1 1/8
10	1	1 7/16

22½° CUT FOR 45° TURN		
SIZE	NO. 1	NO. 2
4	11/16	15/16
6	1	1 3/8
8	1 1/4	1 3/4
10	1 9/16	2 3/16

30° CUT FOR 60° TURN		
SIZE	NO. 1	NO. 2
4	15/16	1 5/16
6	1 5/16	1 7/8
8	1 3/4	2 1/2
10	2 3/16	3 1/16

45° CUT FOR 90° TURN		
SIZE	NO. 1	NO. 2
4	1 9/16	2 1/4
6	2 3/8	3 5/16
8	3 1/16	4 5/16
10	3 13/16	5 3/8

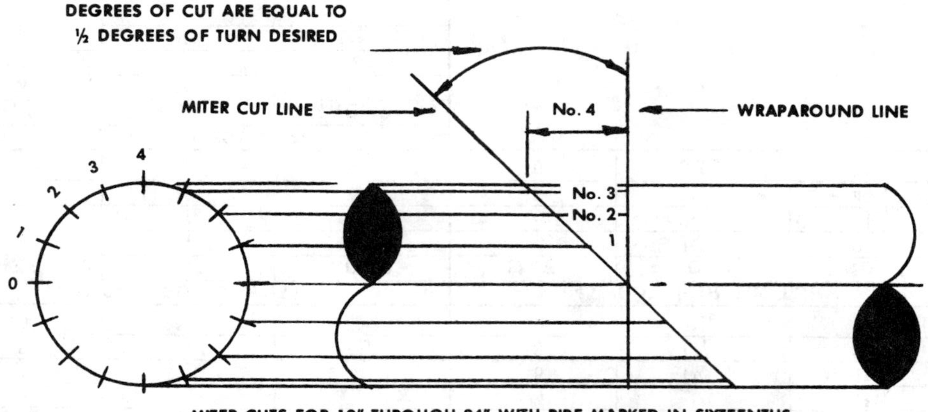

MITER CUTS FOR 12" THROUGH 24" WITH PIPE MARKED IN SIXTEENTHS

LINE No. 4 DIMENSION EQUALS TANGENT OF CUT X O.D. OF PIPE DIVIDED BY 2

LINE No. 3 DIMENSION EQUALS DIMENSION No. 4 X .9239

LINE No. 2 DIMENSION EQUALS DIMENSION No. 4 X .7071

LINE No. 1 DIMENSION EQUALS DIMENSION No. 4 X .3827

12" THROUGH 24" MITER CUTS
MARK PIPE IN SIXTEENTHS

7½° CUT FOR 15° TURN				
SIZE	NO. 1	NO. 2	NO. 3	NO. 4
12	5/16	9/16	3/4	13/16
14	3/8	5/8	7/8	15/16
16	7/16	3/4	1	1 1/16
18	7/16	13/16	1 1/16	1 3/16
20	1/2	15/16	1 3/16	1 5/16
24	5/8	1 1/8	1 7/16	1 9/16
9° CUT FOR 18° TURN				
SIZE	NO. 1	NO. 2	NO. 3	NO. 4
12	3/8	11/16	15/16	1
14	7/16	13/16	1	1 1/8
16	1/2	7/8	1 3/16	1 1/4
18	9/16	1	1 5/16	1 7/16
20	5/8	1 1/8	1 7/16	1 9/16
24	3/4	1 5/16	1 3/4	1 7/8
11¼° CUT FOR 22½° TURN				
SIZE	NO. 1	NO. 2	NO. 3	NO. 4
12	1/2	7/8	1 3/16	1 1/4
14	1/2	1	1 5/16	1 3/8
16	5/8	1 1/8	1 7/16	1 9/16
18	11/16	1 1/4	1 11/16	1 13/16
20	3/4	1 3/8	1 13/16	2
24	15/16	1 11/16	2 3/16	2 3/8

12" THROUGH 24" MITER CUTS
MARK PIPE IN SIXTEENTHS

15° CUT FOR 30° TURN				
SIZE	NO. 1	NO. 2	NO. 3	NO. 4
12	5/8	1 3/16	1 9/16	1 11/16
14	3/4	1 5/16	1 3/4	1 7/8
16	13/16	1 1/2	2	2 1/8
18	15/16	1 11/16	2 1/4	2 3/8
20	1	1 7/8	2 1/2	2 11/16
24	1 1/4	2 1/4	3	3 3/16
22½° CUT FOR 45° TURN				
SIZE	NO. 1	NO. 2	NO. 3	NO. 4
12	1	1 7/8	2 7/16	2 5/8
14	1 1/8	2 1/16	2 11/16	2 7/8
16	1 1/4	2 5/16	3 1/16	3 5/16
18	1 7/16	2 5/8	3 7/16	3 3/4
20	1 9/16	2 15/16	3 13/16	4 1/8
24	1 7/8	3 1/2	4 5/8	5
30° CUT FOR 60° TURN				
SIZE	NO. 1	NO. 2	NO. 3	NO. 4
12	1 3/8	2 5/8	3 3/8	3 11/16
14	1 9/16	2 7/8	3 3/4	4 1/16
16	1 3/4	3 1/4	4 1/4	4 5/8
18	2	3 11/16	4 13/16	5 3/16
20	2 3/16	4 1/16	5 5/16	5 3/4
24	2 5/8	4 7/8	6 3/8	6 15/16

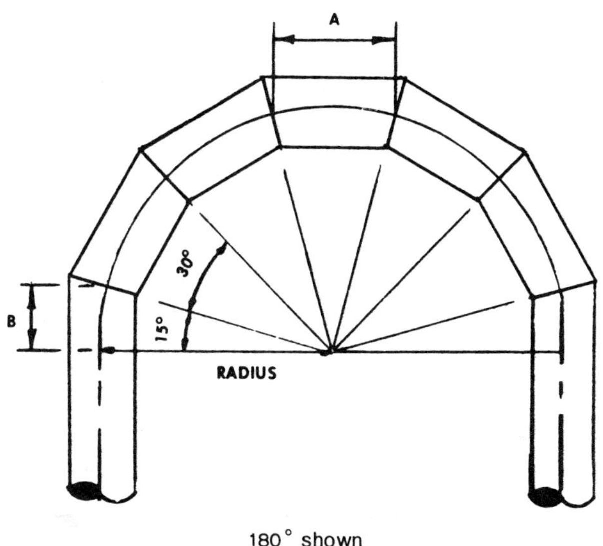

180° shown

FORMULAS FOR MULTIPIECE TURNS TO FORM RADIUS AND NUMBER OF DEGREES DESIRED.

1. ANGLE OF CUT EQUALS
   Degrees of turn divided by (number of welds times 2).

2. LENGTH OF DIMENSION "B"
   Equals radius times Tangent of angle of cut.

3. LENGTH OF PIECES "A" EQUAL
   Dimension "B" times 2.

THREE PIECE 90° TURN
TWO 45° TURNS EQUALS 22½° CUTS

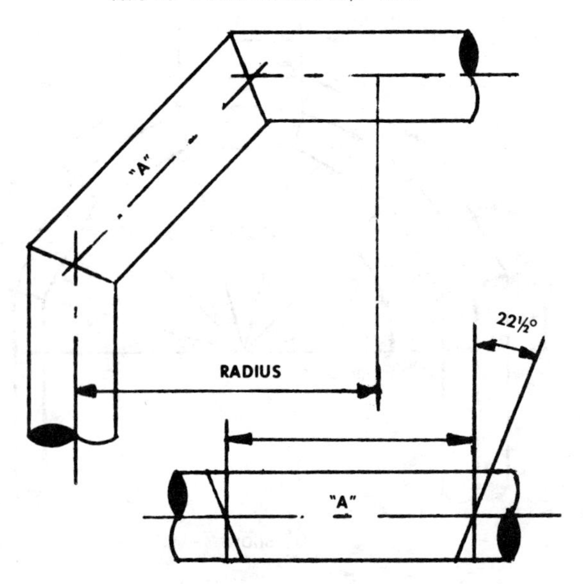

LENGTH "A" EQUALS RADIUS X .414 X 2

RADIUS (Inches)	LENGTH "A" (Inches)
12"	9 13/16"
18	14 3/4
24	19 5/8
30	24 9/16
36	29 7/16
42	34 3/8
48	39 1/4

FOUR PIECE 90° TURN
THREE 30° TURNS W/15° CUTS

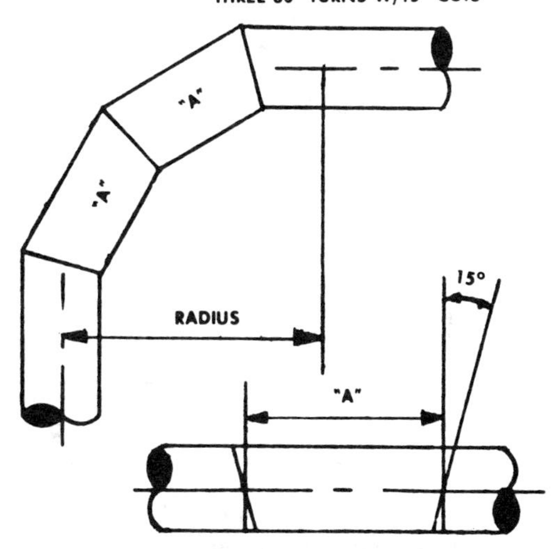

LENGTH "A" EQUALS RADIUS X .2679 X 2

RADIUS (Inches)	LENGTH "A" (Inches)
24"	12⅞"
30	16 1/16
36	19 5/16
42	22½
48	25¾
60	32⅛
72	38 9/16
84	45
96	51 7/16

FIVE PIECE 90° TURN

FOUR 22½° TURNS EQUALS 11¼° CUTS

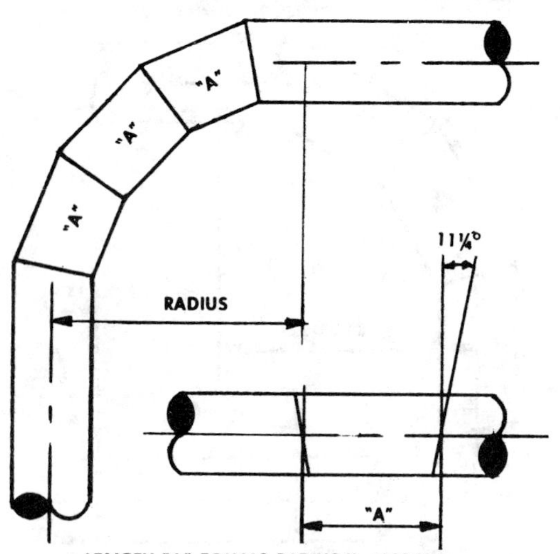

LENGTH "A" EQUALS RADIUS X .1989 X 2

RADIUS (Inches)	LENGTH "A" (Inches)
36"	14 5/16"
42	16 11/16
48	19 1/16
60	23 7/8
72	28 5/8
84	33 7/16
96	38 3/16
108	43
120	47 3/4
132	52 1/2
144	57 1/4

SIX PIECE 90° TURN

FIVE 18° TURNS EQUALS 9° CUTS

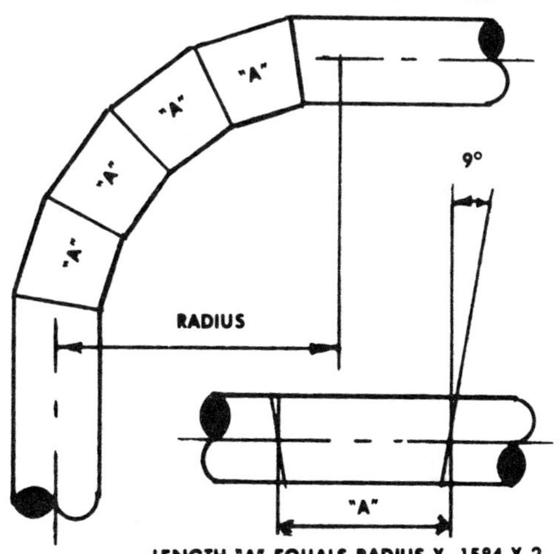

LENGTH "A" EQUALS RADIUS X .1584 X 2

RADIUS (Feet)	LENGTH "A" (Inches)
4'	15 3/16"
5	19
6	22 13/16
7	26 5/8
8	30 3/8
9	34 3/16
10	38
11	41 13/16
12	45 5/8
13	49 3/8
14	53 3/16
15	57

SEVEN PIECE 90° TURN

SIX 15° TURNS EQUALS 7½° CUTS

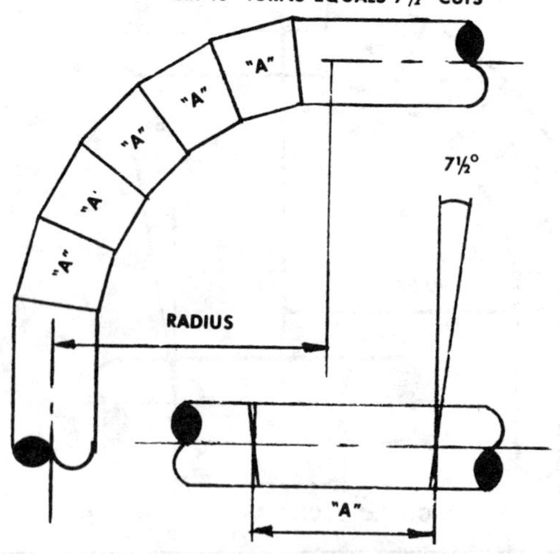

LENGTH "A" EQUALS RADIUS X .1316 X 2

RADIUS (Feet)	LENGTH "A" (Inches)
5'	15 13/16"
6	18 15/16
7	22 1/8
8	25 1/4
9	28 7/16
10	31 9/16
11	34 3/4
12	37 15/16
13	41 1/16
14	44 1/4
15	47 3/8
20	63 3/16

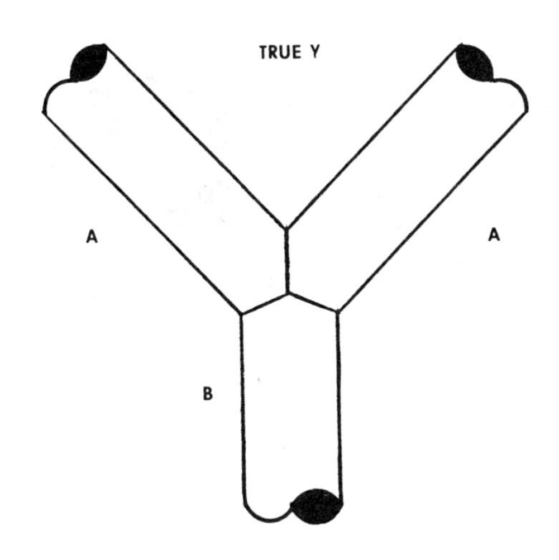

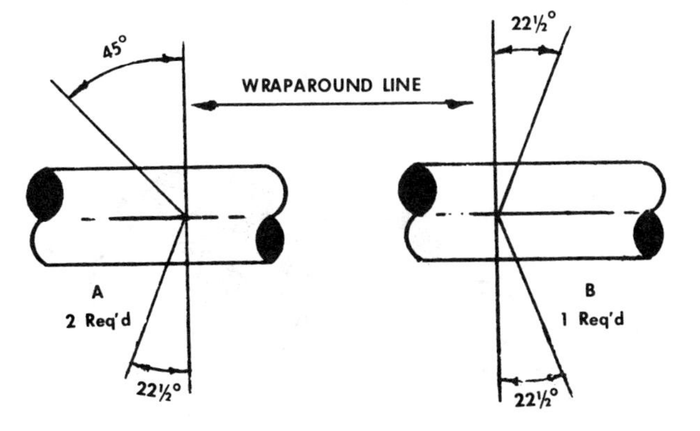

Refer to layout for miter cuts in this book for pipe size to be used.

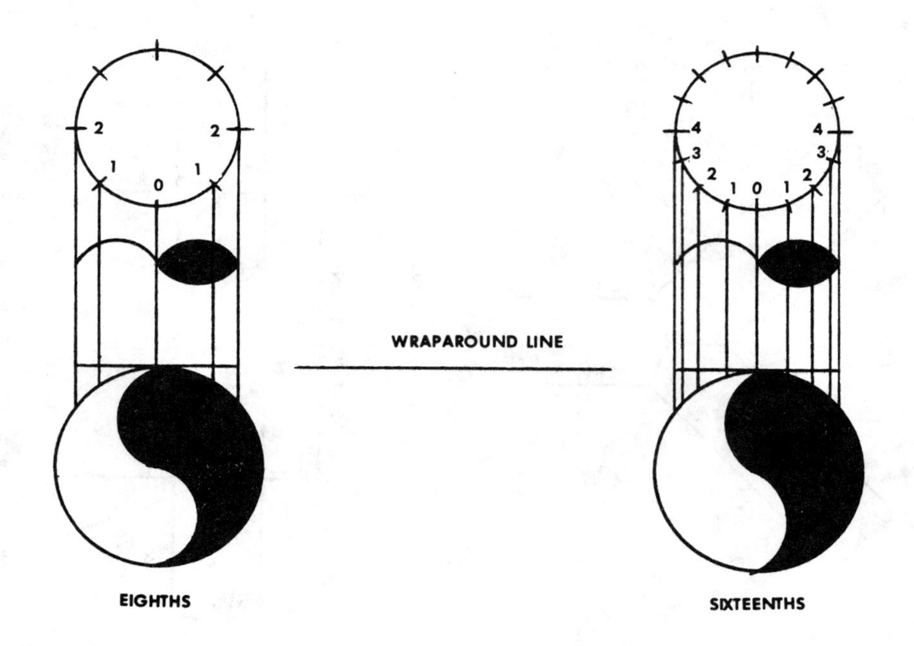
2 2
1 0 1
EIGHTHS
WRAPAROUND LINE
4 4
3 3
2 1 0 1 2
SIXTEENTHS

90° SADDLE ON STANDARD WEIGHT PIPE
PIPE MARKED IN EIGHTHS
SIZE OF HEADER

	3"	4"	6"	8"	10"	12"	14"	16"	18"	20"	22"	24"	NO
3"	3/8	1/4	3/16	1/8	1/8	1/16	1/16	1/16	1/16	1/16	1/16	1/16	1
Riser	15/16	5/8	3/8	5/16	1/4	3/16	3/16	1/8	1/8	1/8	1/8	1/8	2
4"		1/2	5/16	1/4	3/16	3/16	1/8	1/8	1/8	1/8	1/16	1/16	1
Riser		1 1/4	11/16	1/2	3/8	5/16	5/16	1/4	1/4	3/16	3/16	3/16	2
6"			13/16	9/16	7/16	3/8	5/16	5/16	1/4	1/4	3/16	3/16	1
Riser			2	1 1/4	15/16	3/4	11/16	5/8	1/2	1/2	7/16	3/8	2
8"				1 1/16	13/16	11/16	5/8	1/2	7/16	7/16	3/8	5/16	1
Riser				2 11/16	1 3/4	1 3/8	1 1/4	1 1/16	15/16	13/16	3/4	11/16	2
10"					1 5/16	1 1/16	15/16	13/16	3/4	5/8	9/16	9/16	1
Riser					3 7/16	2 7/16	2 1/8	1 3/4	1 1/2	1 3/8	1 3/16	1 1/8	2

90° SADDLE ON STANDARD WEIGHT PIPE PIPE MARKED IN SIXTEENTHS SIZE OF HEADER

	12"	14"	16"	18"	20"	22"	24"	NO
12" Riser	7/16	3/8	5/16	5/16	1/4	1/4	1/4	1
	1 5/8	1 7/16	1 3/16	1 1/16	15/16	7/8	3/4	2
	3 1/4	2 3/4	2 1/4	1 15/16	1 11/16	1 1/2	1 3/8	3
	4 1/4	3 3/8	2 11/16	2 5/16	2	1 3/4	1 5/8	4
14" Riser		1/2	7/16	3/8	5/16	5/16	1/4	1
		1 13/16	1 1/2	1 5/16	1 3/16	1 1/16	15/16	2
		3 5/8	2 7/8	2 3/8	2 1/16	1 7/8	1 11/16	3
		4 3/4	3 1/2	2 15/16	2 1/2	2 1/4	2	4
16" Riser			9/16	1/2	7/16	3/8	3/8	1
			2 1/16	1 13/16	1 9/16	1 7/16	1 1/4	2
			4 3/16	3 3/8	2 7/8	2 9/16	2 5/16	3
			5 9/16	4 1/4	3 1/2	3 1/16	2 3/4	4
18" Riser				5/8	9/16	1/2	7/16	1
				2 3/8	2 1/16	1 7/8	1 11/16	2
				4 13/16	3 15/16	3 7/16	3	3
				6 7/16	4 15/16	4 3/16	3 11/16	4
20" Riser					11/16	5/8	9/16	1
					2 11/16	2 3/8	2 1/8	2
					5 7/16	4 1/2	3 15/16	3
					7 5/16	5 11/16	4 13/16	4
22" Riser						3/4	11/16	1
						2 15/16	2 5/8	2
						6 1/16	5 1/8	3
						8 1/8	6 7/16	4
24" Riser							7/8	1
							3 1/4	2
							6 5/8	3
							9	4

PIPE CIRCUMFERENCE DIVISIONS						
PIPE SIZE	OUT-SIDE DIAM.	CIR.	1/2 CIR.	1/4 CIR.	1/8 CIR.	1/16 CIR.
1 1/2	1.9	5 31/32	3	1 1/2	3/4	3/8
2	2.375	7 15/32	3 3/4	1 7/8	15/16	15/32
2 1/2	2.875	9 1/32	4 1/2	2 1/4	1 1/8	9/16
3	3.5	11	5 1/2	2 3/4	1 3/8	11/16
3 1/2	4	12 9/16	6 9/32	3 1/8	1 9/16	25/32
4	4.5	14 1/8	7 1/16	3 17/32	1 3/4	7/8
5	5.563	17 1/2	8 3/4	4 3/8	2 3/16	1 3/32
6	6.625	20 13/16	10 13/32	5 3/16	2 5/8	1 5/16
8	8.625	27 3/32	13 9/16	6 25/32	3 3/8	1 11/16
10	10.75	33 3/4	16 7/8	8 7/16	4 7/32	2 1/8
12	12.75	40 1/16	20 1/32	10	5	2 1/2
14	14	44	22	11	5 1/2	2 3/4
16	16	50 1/4	25 1/8	12 9/16	6 9/32	3 1/8
18	18	56 9/16	28 9/32	14 1/8	7 1/16	3 17/32
20	20	62 13/16	31 13/32	15 11/16	7 7/8	3 15/16
22	22	69 1/8	34 9/16	17 9/32	8 5/8	4 5/16
24	24	75 13/32	37 11/16	18 27/32	9 7/16	4 23/32
26	26	81 11/16	40 27/32	20 7/16	10 7/32	5 3/32
28	28	87 31/32	44	22	11	5 1/2
30	30	94 1/4	47 1/8	23 9/16	11 25/32	5 7/8
32	32	100 17/32	50 1/4	25 1/8	12 9/16	6 9/32
34	34	106 13/16	53 13/32	26 11/16	13 11/32	6 11/16
36	36	113 3/32	56 9/16	28 9/32	14 1/8	7 1/16
42	42	131 15/16	65 31/32	33	16 1/2	8 1/4
48	48	150 13/16	75 13/32	37 11/16	18 27/32	9 7/16

90° SADDLE ON EXTRA STRONG RISERS MARK IN EIGHTH'S

SIZE OF HEADER

	3"	4"	6"	8"	10"	12"	14"	16"	18"	20"	22"	24"	NO
3"	5/16	1/4	3/16	1/8	1/8	1/16	1/16	1/16	1/16	1/16	1/16	1/16	1
Riser	3/4	1/2	5/16	1/4	3/16	3/16	1/8	1/8	1/8	1/8	1/8	1/16	2
4"		7/16	5/16	3/16	3/16	1/8	1/8	1/8	1/8	1/16	1/16	1/16	1
Riser		1 1/16	5/8	7/16	3/8	5/16	1/4	1/4	3/16	3/16	3/16	1/8	2
6"			11/16	1/2	3/8	5/16	5/16	1/4	1/4	3/16	3/16	3/16	1
Riser			1 11/16	1 1/8	13/16	11/16	5/8	9/16	1/2	7/16	3/8	3/8	2
8"				15/16	3/4	5/8	9/16	7/16	7/16	3/8	5/16	5/16	1
Riser				2 5/16	1 9/16	1 1/4	1 1/8	15/16	7/8	3/4	11/16	5/8	2
10"					1 1/4	1	15/16	3/4	11/16	5/8	9/16	1/2	1
Riser					3 1/8	2 1/4	2	1 11/16	1 7/16	1 1/4	1 1/8	1 1/16	2

90° SADDLE ON EXTRA STRONG RISERS MARK IN SIXTEENTHS

SIZE OF HEADER

	12"	14"	16"	18"	20"	22"	24"	NO
12" Riser	7/16	3/8	5/16	5/16	1/4	1/4	3/16	1
	1 9/16	1 3/8	1 3/16	1	7/8	13/16	3/4	2
	3 1/16	2 9/16	2 1/8	1 13/16	1 5/8	1 7/16	1 5/16	3
	3 7/8	3 3/16	2 9/16	2 3/16	1 15/16	1 11/16	1 9/16	4
14" Riser		7/16	3/8	3/8	5/16	5/16	1/4	1
		1 3/4	1 7/16	1 1/4	1 1/8	1	15/16	2
		3 3/8	2 11/16	2 5/16	2	1 13/16	1 5/8	3
		4 3/8	3 5/16	2 3/4	2 3/8	2 1/8	1 15/16	4
16" Riser			9/16	1/2	7/16	3/8	3/8	1
			2	1 3/4	1 1/2	1 3/8	1 1/4	2
			4	3 1/4	2 13/16	2 7/16	2 3/16	3
			5 3/16	4	3 3/8	2 15/16	2 5/8	4
18" Riser				5/8	9/16	1/2	7/16	1
				2 5/16	2	1 13/16	1 5/8	2
				4 5/8	3 13/16	3 5/16	2 15/16	3
				6 1/16	4 3/4	4	3 1/2	4
20" Riser					11/16	5/8	9/16	1
					2 9/16	2 5/16	2 1/16	2
					5 3/16	4 3/8	3 13/16	3
					6 7/8	5 7/16	4 11/16	4
22" Riser						3/4	11/16	1
						2 7/8	2 9/16	2
						5 13/16	4 15/16	3
						7 3/4	6 3/16	4
24" Riser							13/16	1
							3 3/16	2
							6 7/16	3
							8 9/16	4

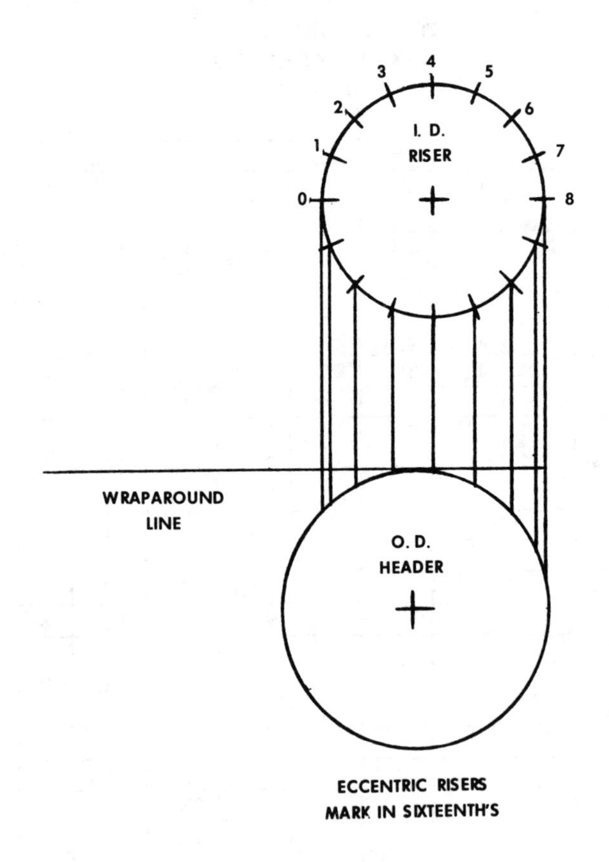

NOTE: These lengths given are for fit up of inside diameters after cuts are made.

90° ECCENTRIC PIPE RISERS
STANDARD WEIGHT RISERS
MARK IN SIXTEENTH'S
SIZE OF HEADER

	4"	6"	8"	10"	12"	14"	16"	18"	20"	22"	24"	NO
3" Riser	1/4	0	1/8	3/8	11/16	1	1 7/16	1 15/16	2 7/16	3	3 9/16	0
	3/16	0	1/8	7/16	3/4	1 1/16	1 1/2	2	2 9/16	3 1/8	3 11/16	1
	1/16	0	1/4	9/16	1	1 1/4	1 3/4	2 5/16	2 7/8	3 7/16	4 1/16	2
	0	1/8	7/16	13/16	1 5/16	1 5/8	2 3/16	2 3/4	3 3/8	4	4 11/16	3
	1/16	3/8	3/4	1 1/4	1 13/16	2 3/16	2 13/16	3 7/16	4 1/8	4 13/16	5 1/2	4
	1/4	3/4	1 1/4	1 7/8	2 1/2	2 7/8	3 5/8	4 5/16	5	5 3/4	6 1/2	5
	5/8	1 1/4	1 7/8	2 1/2	3 1/4	3 11/16	4 7/16	5 1/4	6	6 13/16	7 5/8	6
	1 1/16	1 3/4	2 7/16	3 1/8	3 15/16	4 7/16	5 1/4	6 1/16	6 7/8	7 3/4	8 5/8	7
	1 1/4	2	2 11/16	3 7/16	4 1/4	4 3/4	5 9/16	6 7/16	7 5/16	8 1/8	9	8
4" Riser		1/8	0	1/16	5/16	1/2	7/8	1 1/4	1 11/16	2 3/16	2 11/16	0
		1/8	0	1/8	3/8	9/16	15/16	1 3/8	1 13/16	2 5/16	2 13/16	1
		0	1/16	1/4	9/16	3/4	1 3/16	1 5/8	2 1/8	2 11/16	3 1/4	2
		0	3/16	1/2	7/8	1 1/8	1 5/8	2 1/8	2 11/16	3 5/16	3 7/8	3
		3/16	1/2	15/16	1 3/8	1 3/4	2 5/16	2 7/8	3 1/2	4 3/16	4 13/16	4
		1/2	1	1 9/16	2 1/8	2 1/2	3 3/16	3 7/8	4 9/16	5 1/4	6	5
		1 1/16	1 11/16	2 5/16	3	3 7/16	4 3/16	4 15/16	5 11/16	6 1/2	7 5/16	6
		1 11/16	2 3/8	3 1/16	3 13/16	4 5/16	5 1/8	5 15/16	6 13/16	7 5/8	8 1/2	7
		2	2 11/16	3 7/16	4 1/4	4 3/4	5 9/16	6 7/16	7 5/16	8 1/8	9	8
6" Riser			9/16	1/8	0	0	1/8	3/8	5/8	1	1 3/8	0
			7/16	1/16	0	1/16	3/16	7/16	3/4	1 1/8	1 1/2	1
			3/16	0	1/16	1/8	3/8	11/16	1 1/16	1 7/16	1 7/8	2
			0	1/16	1/4	7/16	3/4	1 3/16	1 5/8	2 1/16	2 9/16	3
			1/8	3/8	3/4	1	1 7/16	1 15/16	2 1/2	3 1/16	3 5/8	4
			9/16	1	1 1/2	1 7/8	2 7/16	3 1/16	3 11/16	4 5/16	5	5
			1 5/16	1 15/16	2 9/16	3	3 11/16	4 3/8	5 1/8	5 7/8	6 5/8	6
			2 3/16	2 15/16	3 11/16	4 1/8	4 15/16	5 3/4	6 9/16	7 3/8	8 1/4	7
			2 11/16	3 7/16	4 1/4	4 3/4	5 9/16	6 7/16	7 5/16	8 1/8	9	8

90 DEGREE ECCENTRIC PIPE RISERS
STANDARD WEIGHT RISERS
MARK IN SIXTEENTH'S
SIZE OF HEADER

	10"	12"	14"	16"	18"	20"	22"	24"	NO
8" Riser	7/8	5/16	1/8	0	0	1/8	5/16	9/16	0
	11/16	1/4	1/16	0	1/16	3/16	3/8	11/16	1
	5/16	1/16	0	1/16	3/16	3/8	11/16	1	2
	0	0	1/16	5/16	9/16	7/8	1 1/4	1 11/16	3
	1/8	5/16	1/2	7/8	1 5/16	1 3/4	2 1/4	2 3/4	4
	5/8	1 1/16	1 3/8	1 7/8	2 7/16	3	3 5/8	4 1/4	5
	1 5/8	2 3/16	2 5/8	3 1/4	3 15/16	4 11/16	5 3/8	6 1/8	6
	2 3/4	3 1/2	4	4 3/4	5 9/16	6 3/8	7 3/16	8	7
	3 7/16	4 1/4	4 3/4	5 9/16	6 7/16	7 5/16	8 1/8	9	8
10" Riser		1 7/16	7/8	3/8	1/8	0	0	1/8	0
		1 1/8	11/16	1/4	1/16	0	1/16	3/16	1
		9/16	1/4	1/16	0	1/16	3/16	3/8	2
		1/16	0	0	3/16	3/8	5/8	15/16	3
		1/16	3/16	7/16	3/4	1 1/8	1 9/16	2	4
		11/16	15/16	1 7/16	1 7/8	2 7/16	3	3 9/16	5
		1 7/8	2 1/4	2 7/8	3 9/16	4 3/16	4 7/8	5 5/8	6
		3 3/8	3 13/16	4 5/8	5 3/8	6 3/16	7	7 13/16	7
		4 1/4	4 3/4	5 9/16	6 7/16	7 5/16	8 1/8	9	8
12" Riser			2 1/2	1 5/16	11/16	5/16	1/16	0	0
			2	1	1/2	3/16	1/16	0	1
			1	7/16	1/8	0	0	1/16	2
			3/16	0	0	1/16	1/4	1/2	3
			0	3/16	3/8	11/16	1	1 3/8	4
			5/8	1	1 7/16	1 15/16	2 7/16	3	5
			2	2 9/16	3 3/16	3 13/16	4 1/2	5 3/16	6
			3 11/16	4 7/16	5 1/4	6	6 13/16	7 5/8	7
			4 3/4	5 9/16	6 7/16	7 5/16	8 1/8	9	8
14" Riser				2 5/16	1 1/4	11/16	5/16	1/8	0
				1 7/8	1	1/2	3/16	1/16	1
				7/8	7/16	1/8	0	0	2
				1/8	0	0	1/8	1/4	3
				1/16	1/4	7/16	3/4	1 1/16	4
				13/16	1 1/4	1 11/16	2 1/8	2 11/16	5
				2 3/8	3	3 5/8	4 1/4	4 15/16	6
				4 3/8	5 1/8	5 7/8	6 11/16	7 1/2	7
				5 9/16	6 7/16	7 5/16	8 1/8	9	8

90 DEGREE ECCENTRIC PIPE RISERS
STANDARD WEIGHT RISERS
MARK IN SIXTEENTH'S
SIZE OF HEADER

	18"	20"	22"	24"	No
16" Riser	2 15/16	1 3/4	1	9/16	0
	2 5/16	1 3/8	3/4	3/8	1
	1 1/8	9/16	1/4	1/16	2
	3/16	1/16	0	1/16	3
	1/16	3/16	7/16	11/16	4
	7/8	1 5/16	1 3/4	2 3/16	5
	2 11/16	3 1/4	3 7/8	4 1/2	6
	4 15/16	5 3/4	6 1/2	7 5/16	7
	6 7/16	7 5/16	8 1/8	9	8
18" Riser		3 1/2	2 1/4	1 3/8	0
		2 13/16	1 3/4	1 1/16	1
		1 3/8	13/16	7/16	2
		1/4	1/16	0	3
		1/16	3/16	3/8	4
		1	1 3/8	1 13/16	5
		2 15/16	3 9/16	4 3/16	6
		5 9/16	6 3/8	7 1/8	7
		7 5/16	8 1/8	9	8
20" Riser			4 3/16	2 3/4	0
			3 5/16	2 3/16	1
			1 11/16	1	2
			5/16	1/8	3
			1/16	3/16	4
			1 1/16	1 7/16	5
			3 1/4	3 7/8	6
			6 3/16	6 15/16	7
			8 1/8	9	8
22" Riser				4 13/16	0
				3 7/8	1
				1 15/16	2
				3/8	3
				1/16	4
				1 1/8	5
				3 9/16	6
				6 13/16	7
				9	8

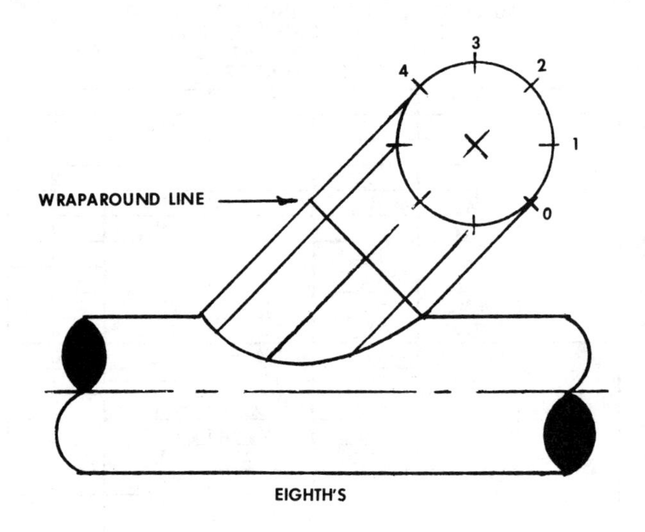

EIGHTH'S

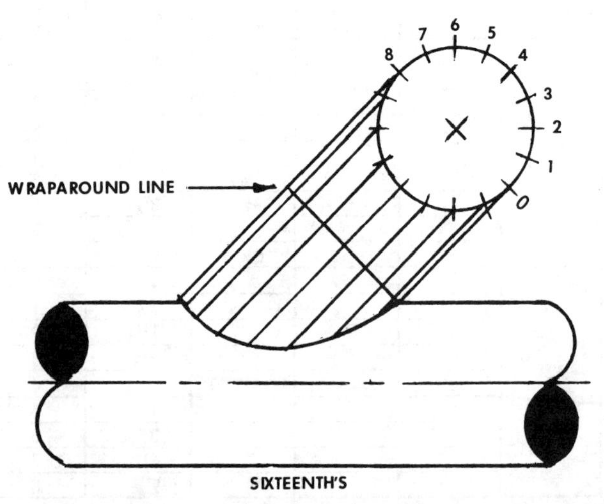

SIXTEENTH'S

45° LATERALS
STANDARD WEIGHT RISERS
MARK IN EIGHTH'S
SIZE OF HEADER

	3"	4"	6"	8"	10"	12"	14"	16"	18"	20"	22"	24"	NO
3" Riser	1	13/16	11/16	5/8	5/8	9/16	9/16	9/16	9/16	9/16	1/2	1/2	1
	2 13/16	2 3/8	2 1/16	1 15/16	1 7/8	1 13/16	1 3/4	1 3/4	1 3/4	1 11/16	1 11/16	1 11/16	2
	3 1/8	3	2 7/8	2 13/16	2 3/4	2 3/4	2 3/4	2 3/4	2 11/16	2 11/16	2 11/16	2 11/16	3
	3 1/16	3 1/16	3 1/16	3 1/16	3 1/16	3 1/16	3 1/16	3 1/16	3 1/16	3 1/16	3 1/16	3 1/16	4
4" Riser		1 5/16	1 1/16	15/16	7/8	13/16	13/16	3/4	3/4	3/4	3/4	11/16	1
		3 3/4	3	2 11/16	2 9/16	2 1/2	2 7/16	2 3/8	2 5/16	2 5/16	2 1/4	2 1/4	2
		4 1/8	3 7/8	3 3/4	3 11/16	3 11/16	3 5/8	3 5/8	3 5/8	3 9/16	3 9/16	3 9/16	3
		4	4	4	4	4	4	4	4	4	4	4	4
6" Riser			2	1 11/16	1 1/2	1 7/16	1 3/8	1 5/16	1 1/4	1 3/16	1 3/16	1 3/16	1
			5 13/16	4 13/16	4 3/8	4 1/8	4	3 7/8	3 3/4	3 11/16	3 5/8	3 9/16	2
			6 5/16	6	5 13/16	5 11/16	5 5/8	5 9/16	5 9/16	5 1/2	5 1/2	5 7/16	3
			6 1/16	6 1/16	6 1/16	6 1/16	6 1/16	6 1/16	6 1/16	6 1/16	6 1/16	6 1/16	4
8" Riser				2 5/8	2 5/16	2 1/8	2	1 7/8	1 13/16	1 3/4	1 11/16	1 5/8	1
				7 3/4	6 1/2	6	5 3/4	5 1/2	5 5/16	5 3/16	5 1/16	4 15/16	2
				8 5/16	7 15/16	7 3/4	7 5/8	7 9/16	7 7/16	7 3/8	7 5/16	7 5/16	3
				8	8	8	8	8	8	8	8	8	4

45° LATERALS
STANDARD WEIGHT RISERS
MARK IN SIXTEENTH'S
SIZE OF HEADER

	10"	12"	14"	16"	18"	20"	22"	24"	NO
	7/8	13/16	3/4	11/16	11/16	5/8	5/8	5/8	1
	3 3/8	3	2 13/16	2 5/8	2 1/2	2 3/8	2 5/16	2 1/4	2
	6 13/16	5 15/16	5 9/16	5 3/16	4 7/8	4 11/16	4 9/16	4 3/8	3
10"	9 7/8	8 7/16	8	7 1/2	7 3/16	6 15/16	6 11/16	6 9/16	4
Riser	10 11/16	9 3/4	9 3/8	9	8 3/4	8 9/16	8 3/8	8 1/4	5
	10 7/16	10 1/16	9 15/16	9 3/4	9 9/16	9 1/2	9 3/8	9 5/16	6
	10 1/8	10 1/16	10	9 15/16	9 15/16	9 7/8	9 7/8	9 7/8	7
	10	10	10	10	10	10	10	10	8
		1 1/16	1	15/16	7/8	13/16	13/16	3/4	1
		4 1/16	3 13/16	3 1/2	3 1/4	3 1/16	2 15/16	2 7/8	2
		8 1/4	7 9/16	6 7/8	6 3/8	6 1/16	5 13/16	5 5/8	3
12"		12	10 13/16	9 13/16	9 1/4	8 13/16	8 1/2	8 1/4	4
Riser		12 7/8	12 1/8	11 7/16	11	10 11/16	10 7/16	10 3/16	5
		12 1/2	12 1/4	11 15/16	11 3/4	11 9/16	11 7/16	11 5/16	6
		12 1/8	12 1/16	12	11 15/16	11 15/16	11 7/8	11 7/8	7
		12	12	12	12	12	12	12	8
			1 3/16	1 1/16	1	15/16	15/16	7/8	1
			4 1/2	4 1/16	3 13/16	3 9/16	3 7/16	3 5/16	2
			9 3/16	8 1/8	7 1/2	7 1/16	6 3/4	6 7/16	3
14"			13 5/16	11 5/8	10 3/4	10 3/16	9 3/4	9 7/16	4
Riser			14 1/4	13 3/16	12 9/16	12 1/8	11 13/16	11 9/16	5
			13 7/8	13 7/16	13 3/16	12 15/16	12 13/16	12 5/8	6
			13 7/16	13 5/16	13 1/4	13 3/16	13 3/16	13 1/8	7
			13 1/4	13 1/4	13 1/4	13 1/4	13 1/4	13 1/4	8
				1 3/8	1 1/4	1 3/16	1 1/8	1 1/16	1
				5 3/16	4 3/4	4 7/16	4 1/4	4 1/16	2
				10 11/16	9 1/2	8 13/16	8 5/16	7 15/16	3
16"				15 1/2	13 9/16	12 5/8	12	11 1/2	4
Riser				16 1/2	15 3/8	14 5/8	14 1/8	13 3/4	5
				16	15 9/16	15 1/4	15	14 13/16	6
				15 7/16	15 3/8	15 5/16	15 1/4	15 3/16	7
				15 1/4	15 1/4	15 1/4	15 1/4	15 1/4	8

45° LATERALS
STANDARD WEIGHT RISERS
MARK IN SIXTEENTH'S
SIZE OF HEADER

	18"	20"	22"	24"	No
18" Riser	1 9/16	1 7/16	1 3/8	1 5/16	1
	5 7/8	5 7/16	5 1/8	4 7/8	2
	12 1/8	10 15/16	10 3/16	9 5/8	3
	17 11/16	15 5/8	14 1/2	13 13/16	4
	18 3/4	17 1/2	16 3/4	16 3/16	5
	18 1/16	17 11/16	17 5/16	17 1/16	6
	17 1/2	17 3/8	17 5/16	17 1/4	7
	17 1/4	17 1/4	17 1/4	17 1/4	8
20" Riser		1 3/4	1 5/8	1 9/16	1
		6 5/8	6 1/8	5 13/16	2
		13 5/8	12 5/16	11 1/2	3
		19 15/16	17 5/8	16 7/16	4
		21	19 11/16	18 7/8	5
		20 3/16	19 3/4	19 7/16	6
		19 1/2	19 7/16	19 5/16	7
		19 1/4	19 1/4	19 1/4	8
22" Riser			1 15/16	1 13/16	1
			7 5/16	6 7/8	2
			15 1/8	13 3/4	3
			22 1/8	19 11/16	4
			23 1/4	21 7/8	5
			22 5/16	21 7/8	6
			21 9/16	21 7/16	7
			21 1/4	21 1/4	8
24" Riser				2 1/8	1
				8	2
				16 9/16	3
				24 3/8	4
				25 1/2	5
				24 7/16	6
				23 9/16	7
				23 1/4	8

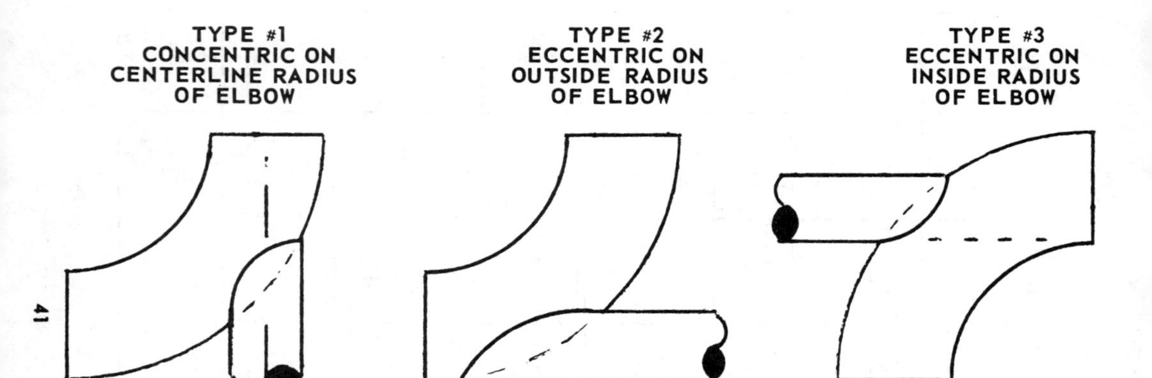

THESE THREE TYPES OF SUPPORTS ON 90° LONG RADIUS WELDELLS REQUIRE DIFFERENT DIMENSIONS.

THE LAYOUT FOR EACH TYPE HOWEVER IS COMMON AND IS SHOWN ON THE FOLLOWING PAGE. TO MARK THE SMALLER SIZES IN EIGHTHS USE ORDINATES #0-2-4-6-8.

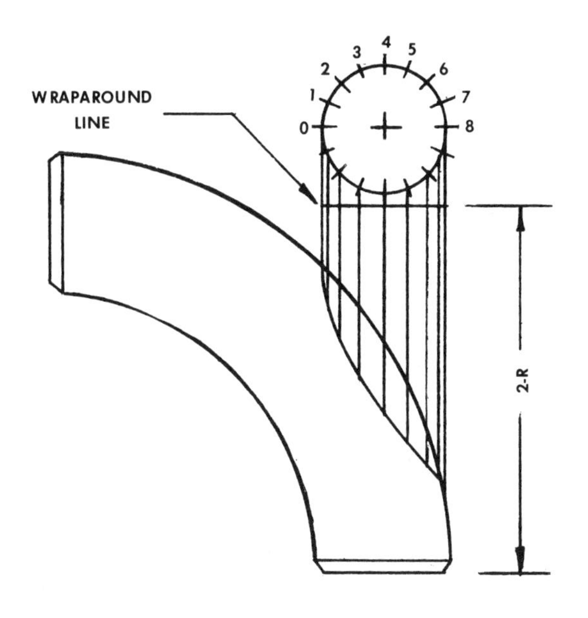

LAYOUT FOR A CONCENTRIC OR AN ECCENTRIC SUPPORT ON BACK OF A 90 DEGREE LONG RADIUS ELBOW

Mark in sixteenth's and measure from a wraparound line that is the length of two radii from the end of the elbow. Example: A 6" long radius elbow has a radius of 9" so your wraparound line will be 18"

CONCENTRIC SUPPORT ON BACK OF 90° L. R. ELBOW (TYPE #1) STANDARD WEIGHT PIPE C/L OF SUPPORT LINES WITH C/L OF ELBOW

	SIZE OF ELBOW							
	2"	3"	4"	6"	8"	10"	12"	No
2" Pipe	2 5/16	3 13/16	5 7/16	8 5/8	11 15/16	15 3/16	18 1/2	0
	2 7/16	3 15/16	5 1/2	8 11/16	12	15 1/4	18 9/16	1
	2 13/16	4 3/16	5 13/16	9	12 5/16	15 1/2	18 7/8	2
	3 3/8	4 11/16	6 1/4	9 3/8	12 11/16	15 15/16	19 1/4	3
	4 1/16	5 3/16	6 11/16	9 13/16	13 1/8	16 3/8	19 11/16	4
	4 1/2	5 9/16	7 1/8	10 1/4	13 9/16	16 13/16	20 1/8	5
	4 3/4	5 7/8	7 7/16	10 9/16	13 7/8	17 1/8	20 7/16	6
	4 7/8	6 1/16	7 5/8	10 13/16	14 1/8	17 3/8	20 11/16	7
	4 7/8	6 1/8	7 11/16	10 7/8	14 3/16	17 7/16	20 3/4	8
3" Pipe		3 1/2	5 1/16	8 3/16	11 1/2	14 11/16	18	0
		3 11/16	5 1/4	8 3/8	11 5/8	14 7/8	18 3/16	1
		4 1/4	5 3/4	8 13/16	12 1/16	15 1/4	18 9/16	2
		5 1/8	6 7/16	9 7/16	12 11/16	15 7/8	19 3/16	3
		6 1/8	7 1/4	10 3/16	13 3/8	16 9/16	19 13/16	4
		6 13/16	7 15/16	10 13/16	14	17 3/16	20 1/2	5
		7 3/16	8 3/8	11 1/4	14 1/2	17 11/16	21	6
		7 5/16	8 9/16	11 9/16	14 13/16	18	21 5/16	7
		7 3/8	8 5/8	11 5/8	14 7/8	18 1/16	21 7/16	8
4" Pipe			4 3/4	7 7/8	11 1/8	14 5/16	17 5/8	0
			5	8 1/16	11 5/16	14 1/2	17 13/16	1
			5 3/4	8 11/16	11 7/8	15 1/16	18 3/8	2
			6 15/16	9 5/8	12 3/4	15 7/8	19 1/8	3
			8 3/8	10 5/8	13 11/16	16 13/16	20 1/16	4
			9 3/8	11 1/2	14 9/16	17 5/8	20 7/8	5
			9 13/16	12 1/16	15 3/16	18 1/4	21 9/16	6
			10	12 3/8	15 1/2	18 11/16	21 15/16	7
			10 1/16	12 1/2	15 5/8	18 13/16	22 1/16	8
6" Pipe				7 1/4	10 3/8	13 1/2	16 3/4	0
				7 5/8	10 3/4	13 13/16	17 1/16	1
				8 3/4	11 3/4	14 3/4	18	2
				10 5/8	13 1/4	16 1/8	19 5/16	3
				12 15/16	14 7/8	17 5/8	20 3/4	4
				14 7/16	16 1/4	18 15/16	22	5
				15 1/16	17 1/16	19 13/16	22 15/16	6
				15 5/16	17 9/16	20 3/8	23 1/2	7
				15 3/8	17 11/16	20 1/2	23 11/16	8

CONCENTRIC SUPPORT ON BACK OF 90° L. R. ELBOW (TYPE #1) STANDARD WEIGHT PIPE C/L OF SUPPORT LINES WITH C/L OF ELBOW

	SIZE OF ELBOW							
	8"	10"	12"	14"	16"	18"	20"	No
8" Pipe	9 13/16	12 7/8	16 1/16	19 3/4	23	26 5/16	29 5/8	0
	10 5/16	13 5/16	16 1/2	20 3/16	23 7/16	26 3/4	30	1
	11 13/16	14 11/16	17 3/4	21 1/2	24 11/16	27 15/16	31 3/16	2
	14 3/8	16 3/4	19 11/16	23 3/8	26 1/2	29 11/16	32 15/16	3
	17 1/2	19	21 3/4	25 7/16	28 1/2	31 5/8	34 13/16	4
	19 9/16	20 13/16	23 1/2	27 3/16	30 1/4	33 3/8	36 9/16	5
	20 3/8	21 7/8	24 5/8	28 7/16	31 1/2	34 5/8	37 7/8	6
	20 11/16	22 7/16	25 5/16	29 1/8	32 1/4	35 7/16	38 5/8	7
	20 3/4	22 5/8	25 1/2	29 3/8	32 1/2	35 11/16	38 15/16	8
10" Pipe		12 1/4	15 3/8	19	22 1/4	25 1/2	28 3/4	0
		12 7/8	16	19 5/8	22 13/16	26 1/16	29 5/16	1
		14 13/16	17 3/4	21 3/8	24 1/2	27 11/16	30 7/8	2
		18	20 7/16	24	26 15/16	30	33 1/8	3
		22 1/8	23 7/16	26 7/8	29 5/8	32 9/16	35 5/8	4
		24 11/16	25 13/16	29 3/16	31 7/8	34 13/16	37 13/16	5
		25 11/16	27 1/8	30 11/16	33 3/8	36 3/8	39 7/16	6
		26 1/16	27 3/4	31 3/8	34 1/4	37 1/4	40 3/8	7
		26 3/16	27 15/16	31 5/8	34 1/2	37 1/2	40 5/8	8
12" Pipe			14 13/16	18 3/8	21 9/16	24 3/4	28	0
			15 9/16	19 1/8	22 1/4	25 7/16	28 11/16	1
			17 7/8	21 3/8	24 3/8	27 1/2	30 5/8	2
			21 13/16	25	27 5/8	30 1/2	33 1/2	3
			26 15/16	29 3/16	31 3/16	33 13/16	36 11/16	4
			30 1/16	32 3/16	34 1/16	36 9/16	39 3/8	5
			31 3/16	33 11/16	35 3/4	38 3/8	41 3/16	6
			31 5/8	34 3/8	36 5/8	39 5/16	42 1/4	7
			31 3/4	34 9/16	36 7/8	39 5/8	42 9/16	8
14" Pipe				18	21 1/8	24 5/16	27 9/16	0
				18 13/16	21 15/16	25 1/8	28 5/16	1
				21 1/2	24 7/16	27 7/16	30 9/16	2
				26 1/16	28 1/4	31	33 7/8	3
				32	32 11/16	34 7/8	37 1/2	4
				35 9/16	35 15/16	38	40 9/16	5
				36 13/16	37 11/16	39 7/8	42 9/16	6
				37 5/16	38 1/2	40 7/8	43 9/16	7
				37 7/16	38 3/4	41 1/8	43 15/16	8

CONCENTRIC SUPPORT ON BACK OF 90° L. R. ELBOW (TYPE #1) STANDARD WEIGHT PIPE C/L OF SUPPORT LINES WITH C/L OF ELBOW

	SIZE OF ELBOW					
	16"	18"	20"	22"	24"	No
16" Pipe	20 1/2	23 11/16	26 13/16	30 1/16	33 5/16	0
	21 1/2	24 5/8	27 3/4	30 15/16	34 3/16	1
	24 9/16	27 1/2	30 1/2	33 9/16	36 11/16	2
	29 7/8	32	34 5/8	37 1/2	40 7/16	3
	36 15/16	37 1/4	39 1/4	41 13/16	44 9/16	4
	41 1/16	41 1/16	42 7/8	45 5/16	48	5
	42 7/16	43	45 1/16	47 9/16	50 5/16	6
	43	43 7/8	46 1/8	48 3/4	51 9/16	7
	43 1/8	44 1/8	46 7/16	49 1/8	52	8
18" Pipe		23 1/16	26 3/16	29 3/8	32 9/16	0
		24 3/16	27 5/16	30 7/16	33 5/8	1
		27 11/16	30 9/16	33 9/16	36 5/8	2
		33 3/4	35 13/16	38 5/16	41 1/8	3
		41 15/16	41 7/8	43 11/16	46 1/8	4
		46 5/8	46 3/16	47 7/8	50 1/8	5
		48 1/8	48 3/8	50 1/4	52 5/8	6
		48 11/16	49 5/16	51 3/8	53 15/16	7
		48 13/16	49 5/8	51 3/4	54 5/16	8
20" Pipe			25 9/16	28 3/4	31 7/8	0
			26 7/8	29 15/16	33 1/16	1
			30 3/4	33 5/8	36 5/8	2
			37 9/16	39 9/16	42 1/16	3
			46 15/16	46 1/2	48 3/16	4
			52 3/16	51 7/16	52 7/8	5
			53 13/16	53 3/4	55 1/2	6
			54 3/8	54 13/16	56 3/4	7
			54 9/16	55 1/16	57 1/16	8
22" Pipe				28 1/8	31 1/4	0
				29 1/2	32 5/8	1
				33 13/16	36 11/16	2
				41 7/16	43 3/8	3
				52	51 3/16	4
				57 13/16	56 11/16	5
				59 1/2	59 3/16	6
				60 1/8	60 5/16	7
				60 1/4	60 9/16	8

ECCENTRIC SUPPORT ON BACK OF 90° L. R. ELBOW (TYPE #2) STANDARD WEIGHT PIPE B. O. P. LINES WITH OUTSIDE RADIUS OF ELBOW

	SIZE OF ELBOW						
	3"	4"	6"	8"	10"	12"	NO
2" Pipe	4 1/4	6 3/8	10 15/16	15 13/16	20 3/4	25 13/16	0
	4 3/8	6 1/2	11 1/8	15 15/16	20 15/16	26 1/16	1
	4 3/4	6 7/8	11 9/16	16 7/16	21 1/2	26 5/8	2
	5 5/16	7 1/2	12 1/4	17 1/4	22 3/8	27 9/16	3
	5 15/16	8 1/4	13 1/8	18 1/4	23 7/16	28 3/4	4
	6 9/16	9	14 1/16	19 5/16	24 11/16	30 1/8	5
	7 1/8	9 11/16	15	20 7/16	25 15/16	31 1/2	6
	7 1/2	10 3/16	15 3/4	21 3/8	27	32 3/4	7
	7 5/8	10 7/16	16 1/16	21 3/4	27 1/2	33 1/4	8
3" Pipe		5 7/16	9 5/8	14 3/16	18 15/16	23 3/4	0
		5 5/8	9 13/16	14 7/16	19 1/8	24 1/16	1
		6 1/8	10 7/16	15 1/16	19 7/8	24 13/16	2
		7	11 5/16	16 1/16	20 15/16	25 15/16	3
		8	12 7/16	17 5/16	22 5/16	27 7/16	4
		8 7/8	13 9/16	18 5/8	23 13/16	29 1/8	5
		9 9/16	14 5/8	19 15/16	25 5/16	30 3/4	6
		10	15 3/8	20 15/16	26 1/2	32 1/8	7
		10 1/8	15 11/16	21 3/8	27 1/16	32 3/4	8
4" Pipe			8 11/16	13	17 1/2	22 1/4	0
			8 15/16	13 1/4	17 13/16	22 9/16	1
			9 11/16	14 1/16	18 11/16	23 7/16	2
			10 13/16	15 5/16	20	24 13/16	3
			12 3/16	16 3/4	21 9/16	26 9/16	4
			13 1/2	18 5/16	23 5/16	28 1/2	5
			14 9/16	19 11/16	25	30 3/8	6
			15 5/16	20 13/16	26 5/16	31 15/16	7
			15 5/8	21 1/4	26 7/8	32 5/8	8
6" Pipe				11 1/16	15 1/4	19 5/8	0
				11 1/2	15 5/8	20	1
				12 5/8	16 13/16	21 1/4	2
				14 5/16	18 9/16	23 1/16	3
				16 3/8	20 11/16	25 5/16	4
				18 1/4	22 13/16	27 11/16	5
				19 11/16	24 11/16	29 7/8	6
				20 5/8	26 1/16	31 9/16	7
				21	26 5/8	32 5/16	8

ECCENTRIC SUPPORT ON BACK OF 90° L. R. ELBOW (TYPE #2) STANDARD WEIGHT PIPE B. O. P. LINES WITH OUTSIDE RADIUS OF ELBOW

	SIZE OF ELBOW						
	10"	12"	14"	16"	18"	20"	NO
8" Pipe	13 9/16	17 11/16	22 1/8	26 1/2	31	35 5/8	0
	14 1/8	18 3/16	22 5/8	27 1/16	31 9/16	36 3/16	1
	15 9/16	19 11/16	24 1/4	28 11/16	33 1/4	37 15/16	2
	17 7/8	22 1/16	26 11/16	31 3/16	35 13/16	40 9/16	3
	20 5/8	24 13/16	29 9/16	34 3/16	39	43 7/8	4
	23 1/16	27 1/2	32 1/2	37 5/16	42 5/16	47 3/8	5
	24 7/8	29 3/4	35 1/16	40 1/4	45 1/2	50 13/16	6
	26	31 7/16	37	42 1/2	48 1/16	53 5/8	7
	26 3/8	32 1/16	37 3/4	43 1/2	49 3/16	54 15/16	8
10" Pipe		16 1/16	20 1/4	24 3/8	28 11/16	33 1/8	0
		16 11/16	20 7/8	25 1/16	29 3/8	33 13/16	1
		18 9/16	22 7/8	27 1/16	31 3/8	35 7/8	2
		21 9/16	25 7/8	30 1/16	34 1/2	39	3
		25	29 7/16	33 11/16	38 3/16	42 13/16	4
		28 1/16	32 11/16	37 3/16	41 7/8	46 3/4	5
		30 1/8	35 3/16	40 1/8	45 1/4	50 7/16	6
		31 3/8	36 7/8	42 5/16	47 13/16	53 5/16	7
		31 13/16	37 1/2	43 3/16	48 7/8	54 5/8	8
12" Pipe			18 3/4	22 3/4	26 13/16	31 1/16	0
			19 9/16	23 1/2	27 5/8	31 7/8	1
			21 15/16	25 7/8	30	34 1/4	2
			25 11/16	29 1/2	33 5/8	37 15/16	3
			30 1/4	33 13/16	37 15/16	42 5/16	4
			33 13/16	37 11/16	42 1/16	46 5/8	5
			35 15/16	40 9/16	45 7/16	50 7/16	6
			37 1/16	42 7/16	47 7/8	53 5/16	7
			37 7/16	43 1/8	48 13/16	54 9/16	8
14" Pipe				21 13/16	25 13/16	29 15/16	0
				22 11/16	26 11/16	30 13/16	1
				25 1/4	29 1/4	33 7/16	2
				29 7/16	33 5/16	37 7/16	3
				34 3/8	38 1/16	42 3/16	4
				38 7/16	42 7/16	46 13/16	5
				41 1/16	45 3/4	50 5/8	6
				42 9/16	47 15/16	53 3/8	7
				43 1/8	48 13/16	54 9/16	8

ECCENTRIC SUPPORT ON BACK OF 90° L. R. ELBOW (TYPE #2) STANDARD WEIGHT PIPE B. O. P. LINES WITH OUTSIDE RADIUS OF ELBOW

	SIZE OF ELBOW				
	18"	20"	22"	24"	No
16" Pipe	24 5/16	28 5/16	32 3/8	36 9/16	0
	25 5/16	29 5/16	33 3/8	37 5/8	1
	28 5/16	32 5/16	36 3/8	40 5/8	2
	33 3/16	36 15/16	41	45 1/4	3
	38 15/16	42 1/2	46 1/2	50 13/16	4
	43 11/16	47 1/2	51 3/4	56 1/4	5
	46 9/16	51 3/16	56	60 15/16	6
	48 1/4	53 5/8	59	64 1/2	7
	48 13/16	54 9/16	60 1/4	66	8
18" Pipe		26 13/16	30 13/16	34 13/16	0
		28	31 15/16	36	1
		31 3/8	35 5/16	39 3/8	2
		36 15/16	40 11/16	44 5/8	3
		43 5/8	47	50 7/8	4
		48 15/16	52 9/16	56 3/4	5
		52 1/8	56 5/8	61 3/8	6
		53 15/16	59 1/4	64 11/16	7
		54 9/16	60 1/4	66	8
20" Pipe			29 3/8	33 5/16	0
			30 5/8	34 9/16	1
			34 7/16	38 3/8	2
			40 11/16	44 3/8	3
			48 5/16	51 1/2	4
			54 1/4	57 3/4	5
			57 11/16	62 1/8	6
			59 5/8	64 15/16	7
			60 1/4	66	8
22" Pipe				31 7/8	0
				33 5/16	1
				37 9/16	2
				44 1/2	3
				53 1/16	4
				59 9/16	5
				63 5/16	6
				65 5/16	7
				66	8

ECCENTRIC SUPPORT ON BACK OF 90° L.R. ELBOW (TYPE #3) STANDARD WEIGHT PIPE B.O.P. LINES WITH INSIDE RADIUS OF ELBOW

	SIZE OF ELBOW						
	3"	4"	6"	8"	10"	12"	No.
	3 7/16"	4 3/4"	7 3/16"	9 11/16"	12 1/8"	14 11/16"	0
	3 9/16	4 13/16	7 1/4	9 3/4	12 3/16	14 11/16	1
2"	3 13/16	5 1/16	7 7/16	9 15/16	12 3/8	14 7/8	2
	4 3/16	5 3/8	7 11/16	10 3/16	12 9/16	15 1/8	3
	4 9/16	5 11/16	8	10 7/16	12 7/8	15 3/8	4
PIPE	4 7/8	6	8 1/4	10 11/16	13 1/16	15 5/8	5
	5 1/16	6 3/16	8 7/16	10 7/8	13 1/4	15 13/16	6
	5 3/16	6 1/4	8 1/2	11	13 3/8	15 7/8	7
	5 3/16	6 5/16	8 9/16	11 1/16	13 7/16	15 15/16	8
		4 3/4"	7 3/16"	9 3/4"	12 3/16"	14 11/16"	0
		4 15/16	7 5/16	9 13/16	12 1/4	14 13/16	1
3"		5 3/8	7 11/16	10 1/8	12 9/16	15 1/16	2
		6	8 1/8	10 9/16	12 15/16	15 7/16	3
		6 11/16	8 11/16	11 1/16	13 3/8	15 7/8	4
PIPE		7 3/16	9 1/16	11 7/16	13 3/4	16 3/16	5
		7 1/2	9 3/8	11 11/16	14	16 1/2	6
		7 5/8	9 1/2	11 7/8	14 3/16	16 11/16	7
		7 11/16	9 9/16	11 15/16	14 1/4	16 11/16	8
			7 3/16"	9 3/4"	12 3/16"	14 11/16"	0
			7 3/8	9 15/16	12 5/16	14 7/8	1
4"			7 15/16	10 3/8	12 3/4	15 1/4	2
			8 11/16	11	13 5/16	15 3/4	3
			9 1/2	11 11/16	13 15/16	16 3/8	4
PIPE			10 1/8	12 1/4	14 7/16	16 7/8	5
			10 1/2	12 5/8	14 13/16	17 1/4	6
			10 11/16	12 13/16	15	17 7/16	7
			10 3/4	12 7/8	15 1/16	17 1/2	8
				9 3/4"	12 3/16"	14 3/4"	0
				10 1/16	12 1/2	15	1
6"				11	13 1/4	15 11/16	2
				12 5/16	14 3/8	16 11/16	3
				13 11/16	15 1/2	17 3/4	4
PIPE				14 3/4	16 3/8	18 9/16	5
				15 5/16	17	19 1/8	6
				15 5/8	17 1/4	19 7/16	7
				15 11/16	17 3/8	19 9/16	8

ECCENTRIC SUPPORT ON BACK OF 90° L.R. ELBOW (TYPE #3) STANDARD WEIGHT PIPE B.O.P. LINES WITH INSIDE RADIUS OF ELBOW

	SIZE OF ELBOW						
	10"	12"	14"	16"	18"	20"	No.
	12 3/16"	14 3/4"	17 15/16"	20 1/2"	23"	25 9/16"	0
	12 5/8	15 1/8	18 5/16	20 13/16	23 3/8	25 7/8	1
8"	13 7/8	16 1/4	19 3/8	21 13/16	24 1/4	26 3/4	2
	15 3/4	17 13/16	20 15/16	23 3/16	25 9/16	28	3
	17 11/16	19 7/16	22 1/2	24 11/16	26 15/16	29 5/16	4
PIPE	19 1/8	20 3/4	23 3/4	25 7/8	28 1/16	30 3/8	5
	19 7/8	21 1/2	24 9/16	26 5/8	28 7/8	31 3/16	6
	20 1/4	21 7/8	24 15/16	27 1/16	29 5/16	31 5/8	7
	20 5/16	21 15/16	25 1/16	27 3/16	29 7/16	31 3/4	8
		14 3/4"	17 15/16"	20 1/2"	23 1/16"	25 9/16"	0
		15 5/16	18 1/2	21	23 1/2	26	1
10"		17	20 1/16	22 7/16	24 13/16	27 1/4	2
		19 7/16	22 3/8	24 1/2	26 11/16	29	3
		22 1/8	24 13/16	26 5/8	28 11/16	30 13/16	4
PIPE		24 1/16	26 5/8	28 1/4	30 3/16	32 5/16	5
		25	27 5/8	29 1/4	31 3/16	33 5/16	6
		25 7/16	28 1/8	29 3/4	31 11/16	33 13/16	7
		25 1/2	28 1/4	29 7/8	31 7/8	34	8
			18"	20 1/2"	23 1/16"	25 9/16"	0
			18 11/16	21 3/16	23 11/16	26 3/16	1
12"			20 7/8	23 1/8	25 7/16	27 13/16	2
			24 5/16	26	28	30 3/16	3
			28 3/16	29 1/8	30 11/16	32 5/8	4
PIPE			30 13/16	31 5/16	32 3/4	34 9/16	5
			32	32 9/16	33 15/16	35 3/4	6
			32 1/2	33 1/16	34 1/2	36 5/16	7
			32 5/8	33 3/16	34 11/16	36 1/2	8
				20 1/2"	23 1/16"	25 9/16"	0
				21 5/16	23 3/4	26 1/4	1
14"				23 5/8	25 7/8	28 3/16	2
				27 1/4	29	31 1/16	3
				31 3/16	32 5/16	34	4
PIPE				33 15/16	34 3/4	36 1/4	5
				35 1/4	36 1/16	37 9/16	6
				35 3/4	36 5/8	38 3/16	7
				35 7/8	36 13/16	38 5/16	8

ECCENTRIC SUPPORT ON BACK OF 90° L.R. ELBOW (TYPE #3) STANDARD WEIGHT PIPE B.O.P. LINES WITH INSIDE RADIUS OF ELBOW

	SIZE OF ELBOW				
	18"	20"	22"	24"	No.
16" PIPE	23 1/16"	25 9/16"	28 1/8"	30 5/8"	0
	23 15/16	26 7/16	28 15/16	31 7/16	1
	26 11/16	28 15/16	31 1/4	33 5/8	2
	31	32 11/16	34 5/8	36 3/4	3
	35 3/4	36 5/8	38 3/16	40	4
	38 15/16	39 1/2	40 13/16	42 1/2	5
	40 7/16	41	42 5/16	44	6
	41	41 5/8	43	44 11/16	7
	41 1/8	41 13/16	43 3/16	44 15/16	8
18" PIPE		25 9/16"	28 1/8"	30 5/8"	0
		26 5/8	29 1/8	31 9/16	1
		29 3/4	32	34 1/4	2
		34 3/4	36 5/16	38 1/4	3
		40 5/16	41	42 3/8	4
		44	44 5/16	45 7/16	5
		45 5/8	46"	47 1/8	6
		46 1/4	46 11/16	47 7/8	7
		46 7/16	46 7/8	48 1/8	8
20" PIPE			28 1/8"	30 5/8"	0
			29 5/16	31 3/4	1
			32 7/8	35 1/16	2
			38 1/2	40 1/16	3
			44 15/16	45 3/8"	4
			49 1/8"	49 3/16	5
			50 7/8	51	6
			51 9/16	51 3/4	7
			51 3/4	52	8
22" PIPE				30 5/8"	0
				32	1
				35 15/16	2
				42 1/4	3
				49 9/16	4
				54 1/4	5
				56 3/16	6
				56 7/8	7
				57 1/16	8

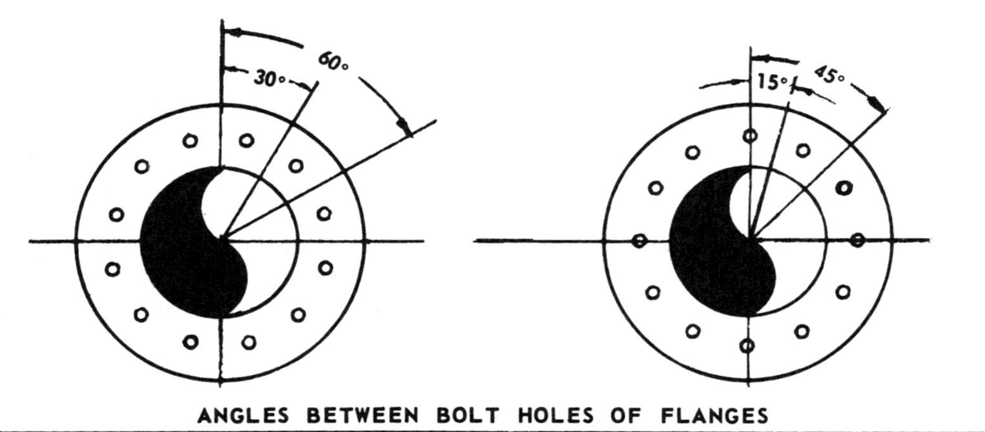

ANGLES BETWEEN BOLT HOLES OF FLANGES

BOLT HOLES STRADDLE C/L	BOLT HOLES ON C/L
4 HOLES 90°	4 HOLES 45°
8 HOLES 45°	8 HOLES 22½°
12 HOLES 30°	12 HOLES 15° – 45°
16 HOLES 22½° – 45°	16 HOLES 11¼° – 33¾°
20 HOLES 18° – 36°	20 HOLES 9° – 27° – 45°
24 HOLES 15° – 30° – 45°	24 HOLES 7½° – 22½° – 37½°

PIPE TEMPLATE LAYOUT USING ORDINATE LENGTHS FROM TABLES IN THIS BOOK.

1. Use a piece of drawing paper or heavier material that is wider than the pipe circumference.

The length should be the dimension of the longest ordinate plus an allowance of 2'' or more for dimension ''A'' shown in drawings. The length of templates for supports on èlbows are an exception and should be the length of the end to center of 2 L. R. elbows as shown.

2. Fit this paper around the pipe and cut it so that the ends of the paper just meet. Be sure that it is kept square with the pipe.

3. Draw the wraparound or reference line and draw the ordinate lines in eighths or sixteenths.

Layout the ordinate lengths from tables in this book. Draw the template curve as shown being sure to contact at least 3 or more points or ordinate lines at all times. A french curve or irregular curve is helpful in drawing so that there will be a smooth curve for better accuracy. The use of sixteen ordinate lines is more accurate than eighths.

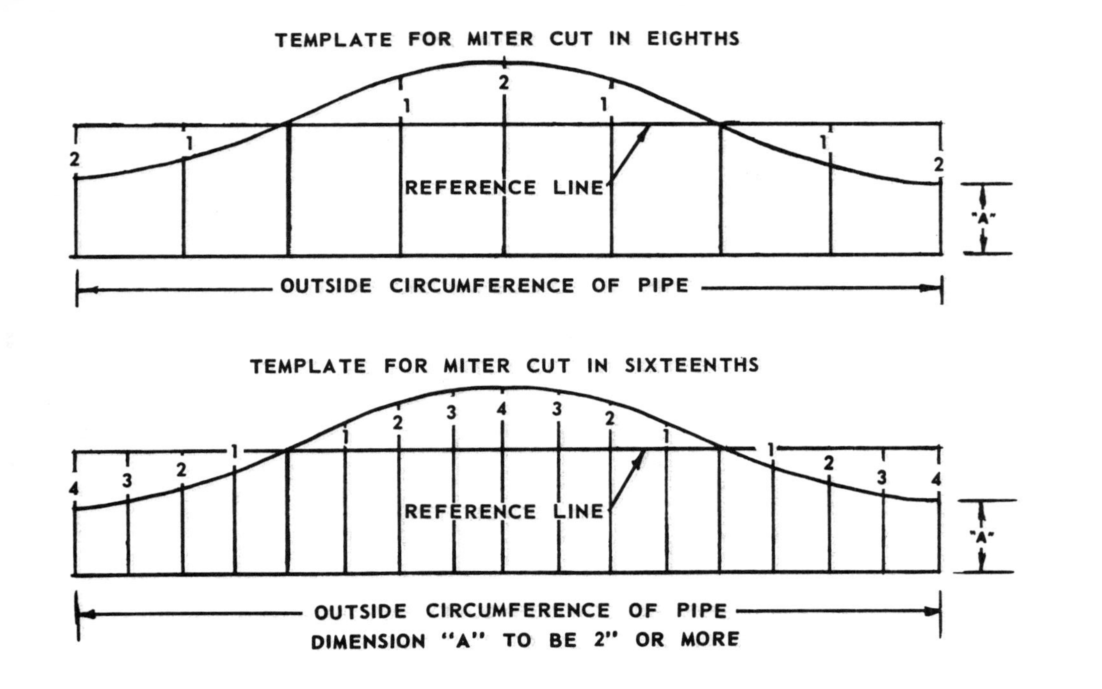
TEMPLATE FOR MITER CUT IN EIGHTHS
2
1
1
2
1
1
2
REFERENCE LINE
"A"
OUTSIDE CIRCUMFERENCE OF PIPE
TEMPLATE FOR MITER CUT IN SIXTEENTHS
4
3
2
1
1
2
3
4
3
2
1
1
2
3
4
REFERENCE LINE
"A"
OUTSIDE CIRCUMFERENCE OF PIPE
DIMENSION "A" TO BE 2" OR MORE

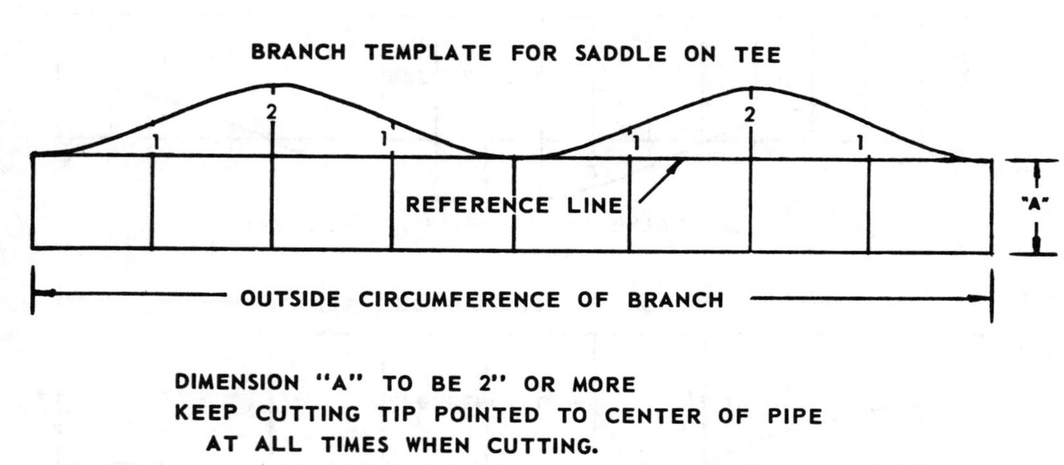

BRANCH TEMPLATE FOR SADDLE ON TEE
1
2
1
1
2
1
REFERENCE LINE
"A"
OUTSIDE CIRCUMFERENCE OF BRANCH
DIMENSION "A" TO BE 2" OR MORE
KEEP CUTTING TIP POINTED TO CENTER OF PIPE
AT ALL TIMES WHEN CUTTING.

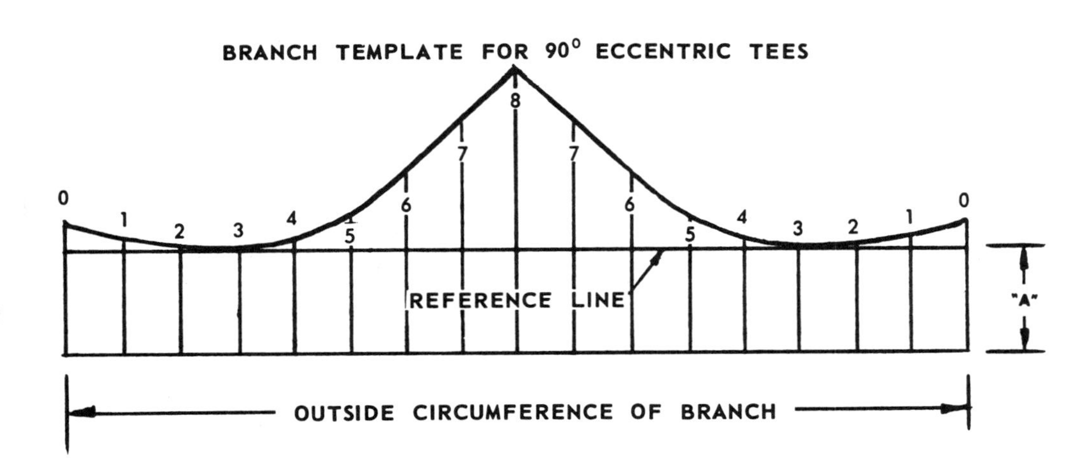
BRANCH TEMPLATE FOR 90° ECCENTRIC TEES
0
1
2
3
4
5
6
7
8
7
6
5
4
3
2
1
0
REFERENCE LINE
"A"
OUTSIDE CIRCUMFERENCE OF BRANCH
DIMENSION "A" TO BE 2" OR MORE
KEEP CUTTING TIP POINTED TO CENTER OF PIPE
AT ALL TIMES WHEN CUTTING.

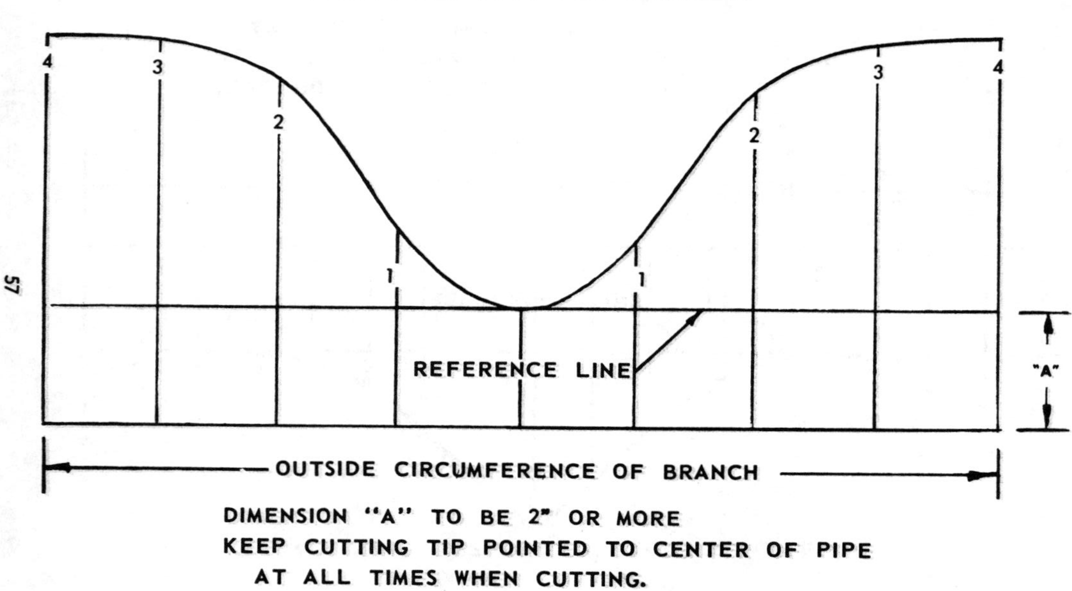
BRANCH TEMPLATE FOR LATERALS
4
3
2
1
1
2
3
4
REFERENCE LINE
"A"
OUTSIDE CIRCUMFERENCE OF BRANCH
DIMENSION "A" TO BE 2" OR MORE
KEEP CUTTING TIP POINTED TO CENTER OF PIPE
AT ALL TIMES WHEN CUTTING.

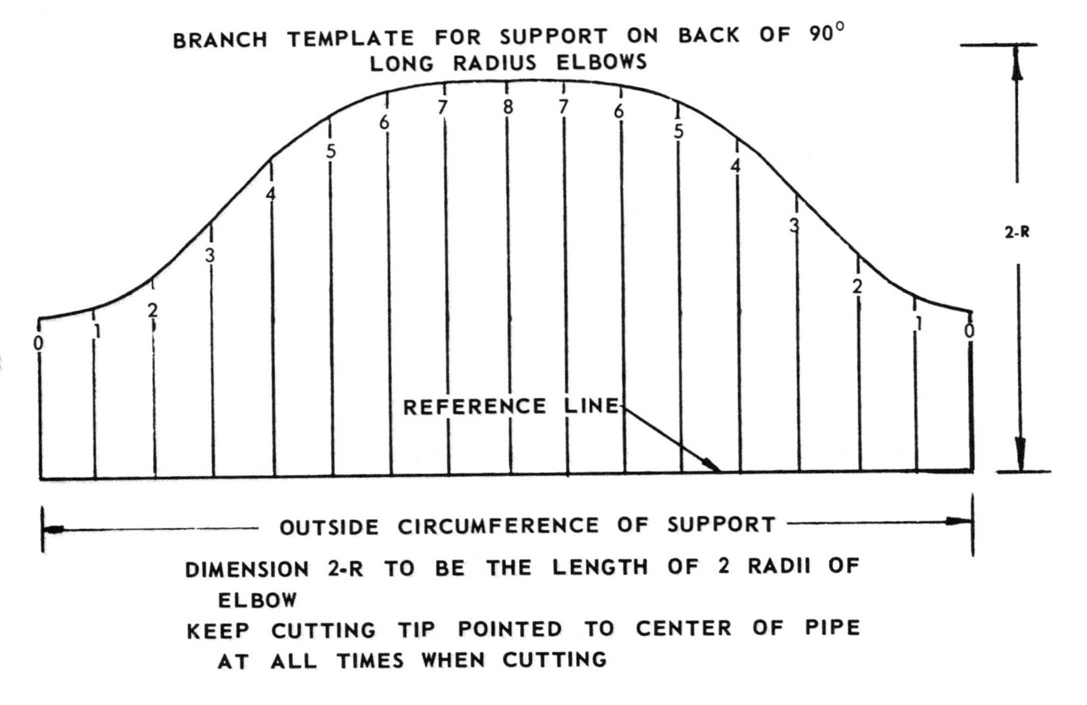
BRANCH TEMPLATE FOR SUPPORT ON BACK OF 90°
LONG RADIUS ELBOWS
0
1
2
3
4
5
6
7
8
7
6
5
4
3
2
1
0
2-R
REFERENCE LINE
OUTSIDE CIRCUMFERENCE OF SUPPORT
DIMENSION 2-R TO BE THE LENGTH OF 2 RADII OF ELBOW
KEEP CUTTING TIP POINTED TO CENTER OF PIPE AT ALL TIMES WHEN CUTTING

1. Draw base line on template.
2. Layout dimensions "L" & ⅓L
3. Layout dimensions "A" for No. of arms
4. Draw centerlines between each arm
5. Layout dimensions "B" & "C"
6. Draw lines to connect points "A", "B" & "C"

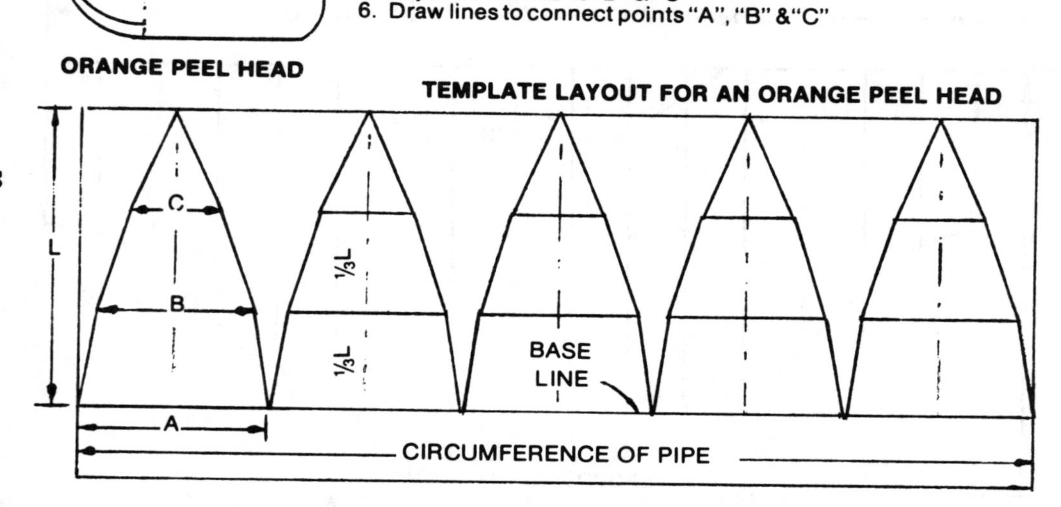

ORANGE PEEL HEAD (Inches)						
Pipe Size	No. of Arms	A	B	C	L	⅓L
2	5	1½	1⁵⁄₁₆	¾	1⅞	⅝
2½	5	1¹³⁄₁₆	1⁹⁄₁₆	⅞	2¼	¾
3	5	2³⁄₁₆	1¹⁵⁄₁₆	1³⁄₃₂	2¾	¹⁵⁄₁₆
3½	5	2½	2³⁄₁₆	1¼	3³⁄₁₆	1¹⁄₁₆
4	5	2¹³⁄₁₆	2¹⁵⁄₃₂	1⁷⁄₁₆	3⁹⁄₁₆	1³⁄₁₆
5	5	3½	3¹⁄₁₆	1¾	4⅜	1⁷⁄₁₆
6	5	4³⁄₁₆	3⅝	2¹⁄₁₆	5¼	1¾
8	6	4½	3¹⁵⁄₁₆	2¼	6¾	2¼
10	7	4¹³⁄₁₆	4⁷⁄₃₂	2¹³⁄₃₂	8⁷⁄₁₆	2¹³⁄₁₆
12	8	5	4⅜	2½	10	3⁵⁄₁₆

FORMULA USED

A = CIRCUMFERENCE OF PIPE O.D. DIVIDED BY NUMBER OF ARMS
B = DIMENSION "A" x .875
C = DIMENSION "A" x .5
L = CIRCUMFERENCE OF PIPE O.D. DIVIDED BY 4

NUMBER OF ARMS = CIRCUMFERENCE OF PIPE O.D. DIVIDED BY 5
FIVE ARMS TO BE MINIMUM

USE A RADIAL CUT.

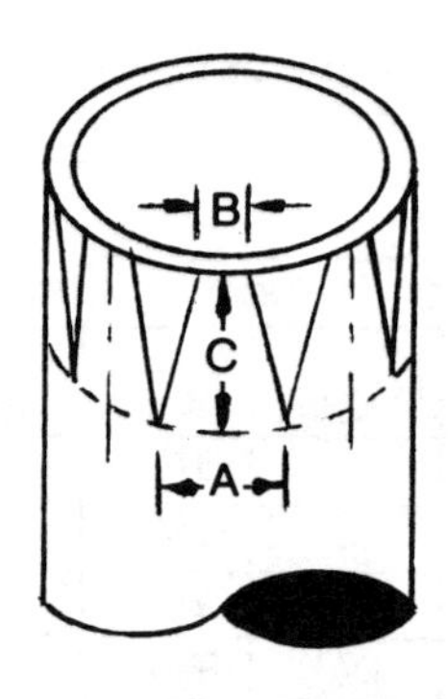

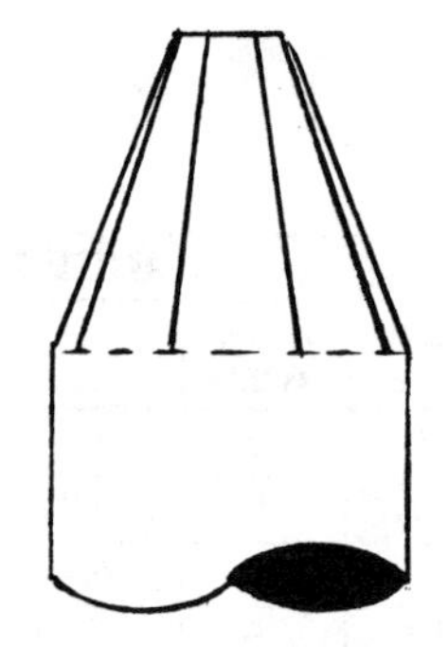

CONCENTRIC REDUCER LAYOUT

"A" = Circumference of large pipe divided by No. of arms.

"B" = Circumference of small pipe divided by No. of arms

"C" = Difference of pipe O. D.'s x 1.3

"N" = Number of arms = difference of pipe O. D.'s x 1.33 Minimum 4 arms.

1. Draw wraparound line on pipe equal to dimension "C".
2. On this line mark off divisions per dimension "A".
3. Draw lines on pipe halfway between each dimension "A"
4. Mark off dimension "B" on these lines at end of pipe with ½ on each side of line.
5. Draw lines to connect points "B" to points at wraparound line.
6. Burn out sections between arms using a radial cut, then bevel arms.

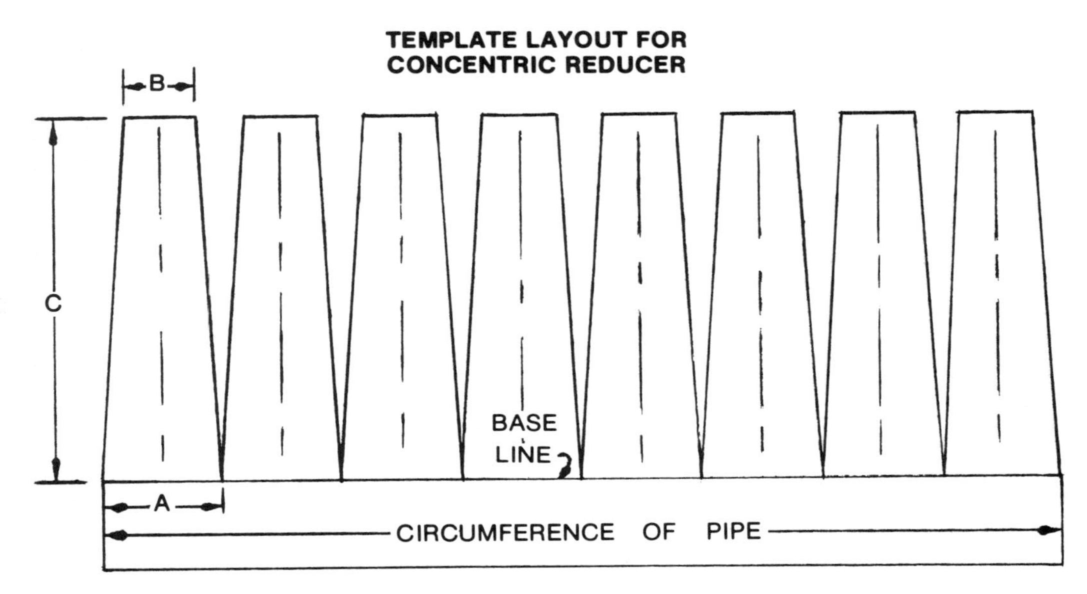
TEMPLATE LAYOUT FOR
CONCENTRIC REDUCER
B
C
BASE LINE
A
CIRCUMFERENCE OF PIPE

CONCENTRIC REDUCERS (Inches)

Pipe Size	No. of Arms	A	B	C	Pipe Size	No. of Arms	A	B	C
2 x 1½	5	1½	1 3/16	5/8	4 x 3½	5	2 13/16	2½	11/16
2 x 1¼	5	1½	1 1/16	15/16	4 x 3	5	2 13/16	2 3/16	1 5/16
2 x 1	5	1½	13/16	1 3/8	4 x 2½	5	2 13/16	1 13/16	2 1/8
2½ x 2	5	1 13/16	1½	11/16	4 x 2	5	2 13/16	1½	2¾
2½ x 1½	5	1 13/16	1 3/16	1¼	4 x 1½	5	2 13/16	1 3/16	3 3/8
2½ x 1¼	5	1 13/16	1 1/16	1 9/16	4 x 1¼	5	2 13/16	1 1/16	3 11/16
2½ x 1	5	1 13/16	13/16	2	4 x 1	5	2 13/16	13/16	4 1/8
3 x 2½	5	2 3/16	1 13/16	13/16	5 x 4	5	3½	2 13/16	1 3/8
3 x 2	5	2 3/16	1½	1½	5 x 3½	5	3½	2½	2 1/16
3 x 1½	5	2 3/16	1 3/16	2 1/8	5 x 3	5	3½	2 3/16	2 11/16
3 x 1¼	5	2 3/16	1 1/16	2 3/8	5 x 2½	5	3½	1 13/16	3½
3 x1	5	2 3/16	13/16	2 7/8	5 x 2	5	3½	1½	4 1/8
3½ x 3	5	2½	2 3/16	11/16	5 x 1½	5	3½	1 3/16	4¾
3½ x 2½	5	2½	1 13/16	1½	5 x 1¼	5	3½	1 1/16	5 1/16
3½ x 2	5	2½	1½	2 1/8	5 x 1	5	3½	13/16	5½
3½ x 1½	5	2½	1 3/16	2¾					
3½ x 1¼	5	2½	1 1/16	3 1/16					
3½ x 1	5	2½	13/16	3½					

CONCENTRIC REDUCERS (Inches)

Pipe Size	No. of Arms	A	B	C	Pipe Size	No. of Arms	A	B	C
6 x 5	5	4 5/32	3 1/2	1 3/8	10 x 8	7	4 13/16	3 7/8	2 3/4
6 x 4	5	4 5/32	2 13/16	2 3/4	10 x 6	7	4 13/16	3	5 3/8
6 x 3 1/2	5	4 5/32	2 1/2	3 7/16	10 x 5	7	4 13/16	2 1/2	6 3/4
6 x 3	5	4 5/32	2 3/16	4 1/16	10 x 4	8	4 7/32	1 3/4	8 1/8
6 x 2 1/2	5	4 5/32	1 13/16	4 7/8	10 x 3 1/2	9	3 3/4	1 3/8	8 3/4
6 x 2	6	3 15/32	1 1/4	5 1/2	10 x 3	10	3 3/8	1 1/8	9 7/16
6 x 1 1/2	6	3 15/32	1	6 1/8	10 x 2 1/2	11	3 1/16	13/16	10 1/4
8 x 6	6	4 1/2	3 15/32	2 5/8	12 x 10	8	5	4 7/32	2 5/8
8 x 5	6	4 1/2	2 29/32	4	12 x 8	10	4	2 23/32	5 3/8
8 x 4	6	4 1/2	2 11/32	5 3/8	12 x 6	10	4	2 3/32	8
8 x 3 1/2	6	4 1/2	2 3/32	6	12 x 5	10	4	1 3/4	9 3/8
8 x 3	7	3 7/8	1 9/16	6 11/16	12 x 4	11	3 21/32	1 9/32	10 3/4
8 x 2 1/2	8	3 3/8	1 1/8	7 1/2	12 x 3 1/2	12	3 11/32	1 1/16	11 3/8
8 x 2	8	3 3/8	15/16	8 1/8	12 x 3	12	3 11/32	15/16	12

TEMPLATE LAYOUT FOR AN ECCENTRIC REDUCER

The use of a template for an eccentric reducer is simpler and more accurate than marking off the pipe. Use sheet metal or gasket material about 1/32" thick. The material should be slightly longer than the pipe circumference. The width should be about 4" more than dimension "E". Check the material and be sure it is exactly square. Fit the material around the circumference of the pipe and mark and cut it so that the ends of material just meet on the pipe. The steps below should be followed.

1. Draw a base line on the template 3" up from the edge.
2. Draw a center line on the template in the exact center for arm #1.
3. At this centerline on the base line mark off 1/8 circumferences on each side of template.

START THESE STEPS AT THE LEFT SIDE OF TEMPLATE AND REPEAT EACH STEP FOR THE RIGHT SIDE.

1. At left edge of template mark dimension "F" from base line.
2. At this point mark off dimension 1½ A for arm #4.
3. Mark off dimension "B".
4. Mark off dimension "A" for arm #3.
5. Mark off dimension "C".
6. Mark off dimension "A" for arm #2.
7. At centerline of template mark off ½ of dimension "A" on each side for arm #1.
8. Draw in lines for each arm to points marked on base line.

   Use a radial cut and bevel each arm after cutting. Heat and shape the bottom arm first, then heat the remaining arms so that they may be pulled down as well as in.

   These eccentric reducers can be cut back for each larger size of pipe as required.

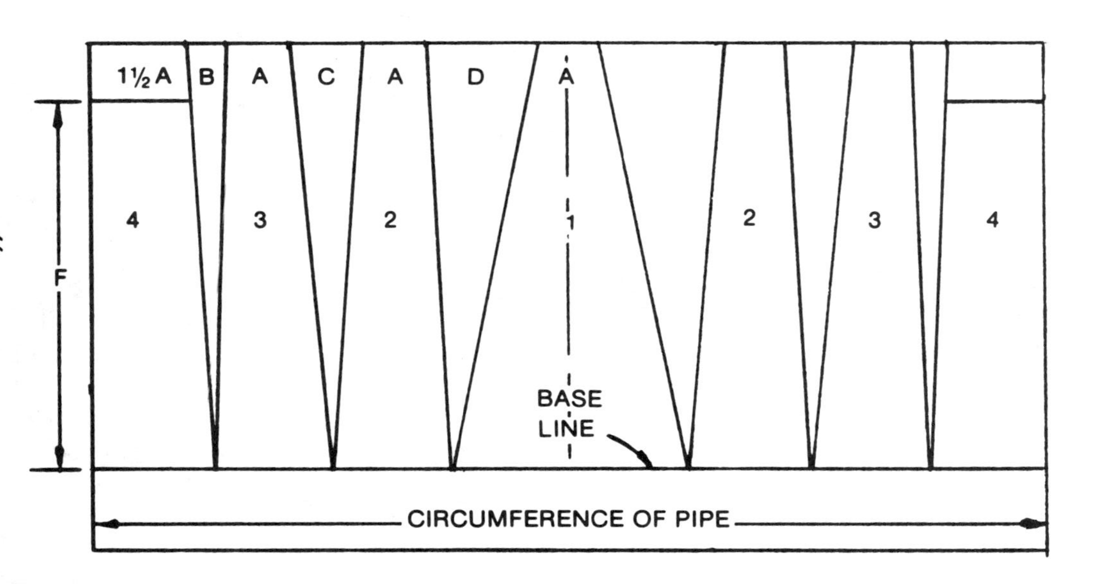
TEMPLATE LAYOUT
FOR ECCENTRIC REDUCERS
1½ A
B
A
C
A
D
A
4
3
2
1
2
3
4
F
BASE
LINE
CIRCUMFERENCE OF PIPE

ECCENTRIC REDUCER

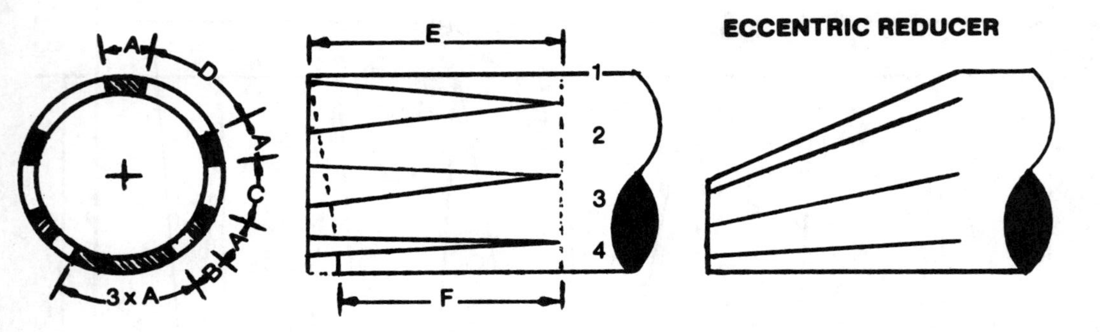

FORMULA

"A" = ⅛ of small pipe circumference.
"B" = Difference of outside circumferences x .0833
"C" = Difference of outside circumferences x .1666
"D" = Difference of outside circumferences x .25
"E" = 1½ x O.D. of larger pipe
"F" = Dimension "E" x .866

Use a radial cut.

ECCENTRIC REDUCERS (Inches)			
3 x 2		3½ x 2	
A =	$^{15}/_{16}$	A =	$^{15}/_{16}$
B =	$^{5}/_{16}$	B =	$^{7}/_{16}$
C =	$^{9}/_{16}$	C =	⅞
D =	⅞	D =	1¼
E =	5¼	E =	6
F =	4$^{9}/_{16}$	F =	5$^{3}/_{16}$
1½ A =	1$^{13}/_{32}$	1½ A =	1$^{13}/_{32}$
3″ Pipe Circum =	11″	3½″ Pipe Circum=	12$^{9}/_{16}$″
⅛ Circum =	1⅜″	⅛ Circum =	1$^{9}/_{16}$″

ECCENTRIC REDUCERS (Inches)			
4 x 2		5 x 2½	
A =	$^{15}/_{16}$	A =	1⅛
B =	$^{9}/_{16}$	B =	$^{11}/_{16}$
C =	1⅛	C =	1⅜
D =	1$^{11}/_{16}$	D =	2⅛
E =	6¾	E =	8⅜
F =	5⅞	F =	7¼
1½ A =	1$^{13}/_{32}$	1½ A =	1$^{11}/_{16}$
4″ Pipe Circum =	14⅛″	5″ Pipe Circum =	17½″
⅛ Circum =	1¾″	⅛ Circum =	2$^{3}/_{16}$″

ECCENTRIC REDUCER (Inches)			
6 x 3		8 x 4	
A =	$1\frac{3}{8}$	A =	$1\frac{3}{4}$
B =	$\frac{13}{16}$	B =	$1\frac{1}{16}$
C =	$1\frac{5}{8}$	C =	$2\frac{1}{8}$
D =	$2\frac{7}{16}$	D =	$3\frac{1}{4}$
E =	10	E =	13
F =	$8\frac{11}{16}$	F =	$11\frac{1}{4}$
$1\frac{1}{2}$ A =	$2\frac{1}{16}$	$1\frac{1}{2}$ A =	$2\frac{5}{8}$
6" Pipe Circum =	$20\frac{13}{16}$"	8" Pipe Circum =	$27\frac{1}{8}$"
$\frac{1}{8}$ Circum =	$2\frac{5}{8}$"	$\frac{1}{8}$ Circum =	$3\frac{3}{8}$"

ECCENTRIC REDUCERS (Inches)			
10 x 6		12 x 6	
A =	$2\frac{5}{8}$	A =	$2\frac{5}{8}$
B =	$1\frac{1}{16}$	B =	$1\frac{5}{8}$
C =	$2\frac{1}{8}$	C =	$3\frac{3}{16}$
D =	$3\frac{1}{4}$	D =	$4\frac{13}{16}$
E =	$16\frac{1}{8}$	E =	$19\frac{1}{8}$
F =	14	F =	$16\frac{9}{16}$
$1\frac{1}{2}$ A =	$3\frac{15}{16}$	$1\frac{1}{2}$ A =	$3\frac{15}{16}$
10" Pipe Circum =	$33\frac{3}{4}$"	12" Pipe Circum =	$40\frac{1}{16}$"
$\frac{1}{8}$ Circum =	$4\frac{7}{32}$"	$\frac{1}{8}$ Circum =	5"

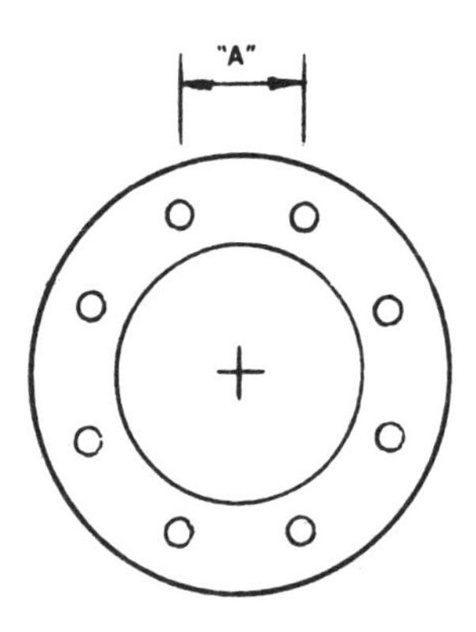

LAYING OUT HOLES IN FLANGES

FORMULA: For finding Dim. "A", multiply the bolt circle diameter times the SINE of one half of the angle between the holes. 45° minus 22½° shown.

NO OF HOLES	BOLT CIRCLE DIAM X	NO OF HOLES	BOLT CIRCLE DIAM X
4	.7071	20	.1564
6	.500	24	.1305
8	.3827	28	.1120
12	.2588	32	.0980
16	.1951	36	.0871

HOW TO LAY OUT ORDINATE LINES AND LENGTHS FOR A CONCENTRIC OR A TANGENTIAL NOZZLE

1. Set a pair of dividers to a radius that will equal the I.D. of the nozzle when it is to be fitted to the outside wall of the vessel. Set dividers for the O.D. of the nozzle if it is to fit the inside wall of the vessel.
2. With dividers correctly set scribe an arc of 180° on a piece of gasket material or sheet metal and draw a line across this half circle.
3. Draw lines #0 and #4 the length of this material.
4. Use dividers to step off each half of the semi-circle into 4 equal sections of 22½°. At these points on half circle draw the lines #1, #2, and #3 as before.
5. On a table or other surface scribe an arc at a radius that will equal the O.D. or the I.D. of the vessel wall that you will fit the nozzle to.
6. Place marked off material in exact position you want on this arc and make sure it is square with the vessel. If you are making a tangential type nozzle be sure that the O.D. of the nozzle does not extend beyond the outside wall of the vessel.
7. Hold material in position and at high point of vessel wall draw a reference or wraparound line onto the material. Sometimes line #0 has a length so be sure you have the high point.
8. Scribe the vessel radius onto the material. You now have the ordinates lengths on the material.

   The template layout for either of these types is shown in the template layout section of this book.

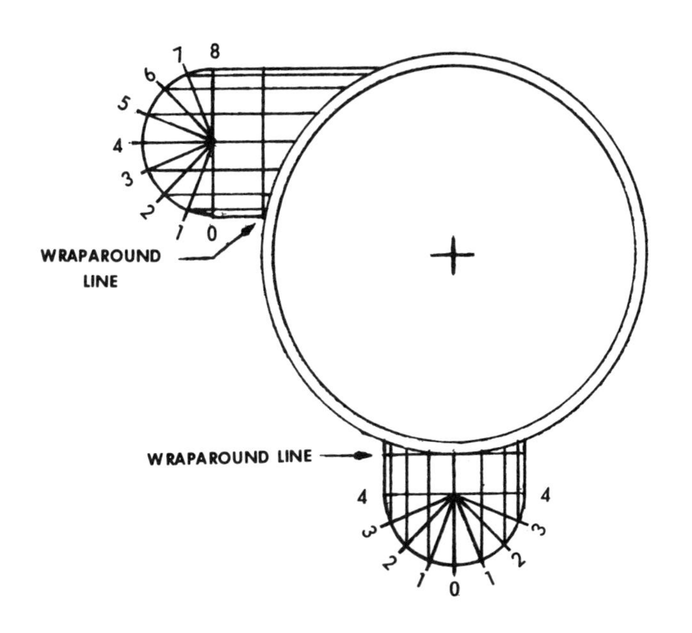

ORDINATE LINES AND LENGTHS

Spacing ordinate lines with dividers is recommended, however these spacings can also be calculated. Line numbers below are from concentric type.

Line #4 = ½ the I.D. or the O.D. of the pipe or nozzle.
Line #1 = Dimension #4 × .3827
Line #2 = Dimension #4 × .707
Line #3 = Dimension #4 × .9239

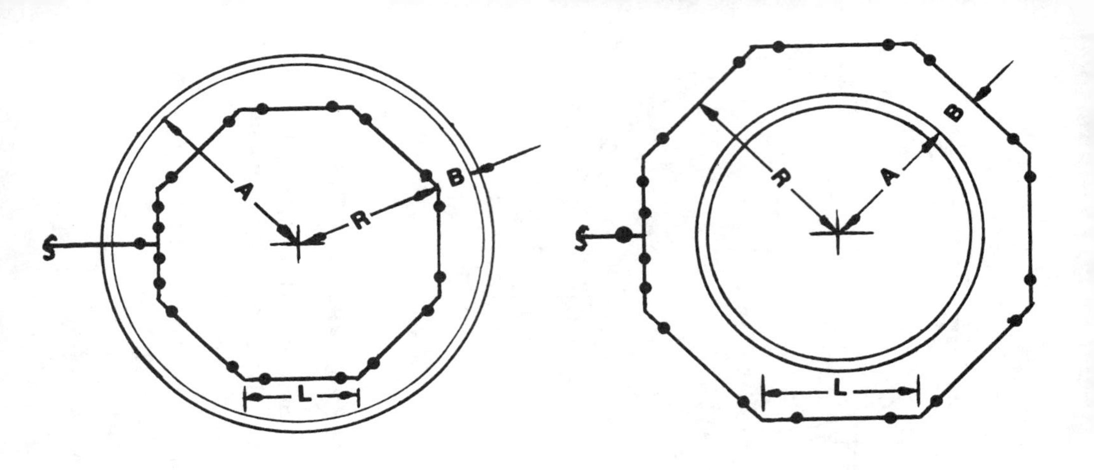

PIPE COIL INSIDE TANK

FORMULA FOR DIMENSION "L"
"R" × 2 × SINE of ½
degrees of sector.

PIPE COIL OUTSIDE TANK

FORMULA FOR DIMENSION "L"
"R" × 2 × TANGENT of ½
degrees of sector.

TANK COILS

COIL INSIDE TANK			COIL OUTSIDE TANK		
A = Inside radius of tank **B** = Clearance inside tank **R** = Radius of coil **L** = Center to center length			**A** = Outside radius of tank **B** = Clearance outside tank **R** = Radius of coil **L** = Center to center length		
Angle of Fitting	**No. of Pipes per Coil**	**Sine**	**Angle of Fitting**	**No. of Pipes per Coil**	**Tangent**
90°	4	.707	90°	4	1.000
60°	6	.500	60°	6	.577
45°	8	.3827	45°	8	.414
30°	12	.2588	30°	12	.2679
22½°	16	.195	22½°	16	.1989
11¼°	32	.098	11¼°	32	.0985

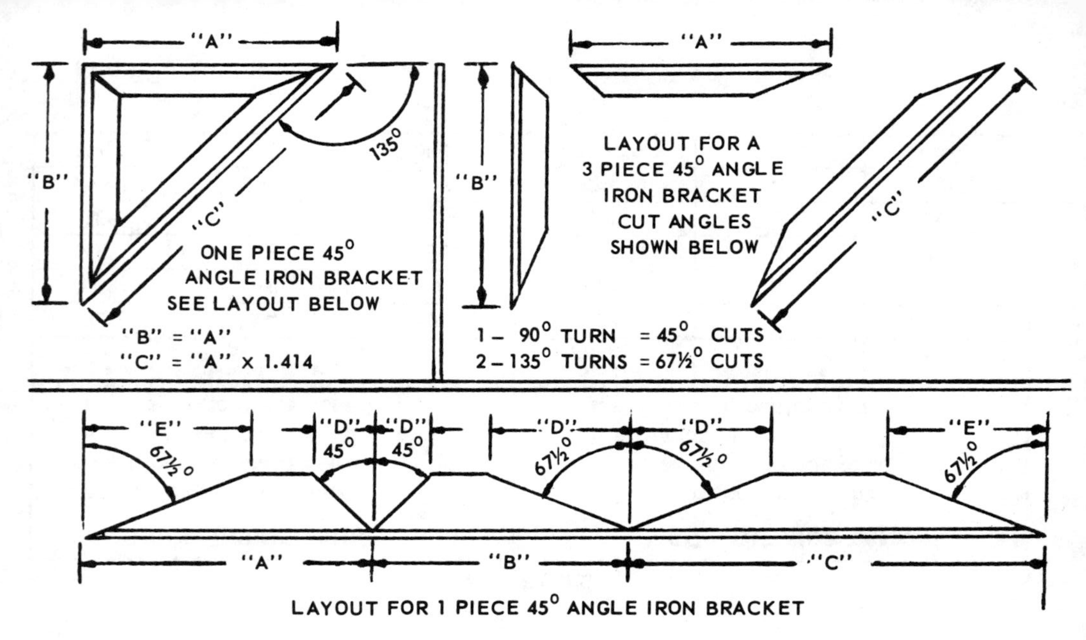
"A"
"B"
"C"
135°
ONE PIECE 45°
ANGLE IRON BRACKET
SEE LAYOUT BELOW
"B" = "A"
"C" = "A" x 1.414
LAYOUT FOR A
3 PIECE 45° ANGLE
IRON BRACKET
CUT ANGLES
SHOWN BELOW
1 – 90° TURN = 45° CUTS
2 – 135° TURNS = 67½° CUTS
"E"
"D"
45°
67½°
LAYOUT FOR 1 PIECE 45° ANGLE IRON BRACKET

DIMENSIONS FOR THE LAYOUT OF ANGLE IRON MITER CUTS

11¼° CUT FOR 22½° TURN = WIDTH × .1989			30° CUT FOR 60° TURN = WIDTH × .5773		
Size	One Piece "D"	Two Piece "E"	Size	One Piece "D"	Two Piece "E"
1/8" × 1"	3/16"	3/16"	1/8" × 1"	1/2"	9/16"
1/4 × 1 1/2	1/4	5/16	1/4 × 1 1/2	3/4	7/8
1/4 × 2	5/16	3/8	1/4 × 2	1	1 1/8
3/8 × 2 1/2	7/16	1/2	3/8 × 2 1/2	1 1/4	1 7/16
3/8 × 3	1/2	9/16	3/8 × 3	1 1/2	1 3/4
3/8 × 4	3/4	13/16	3/8 × 4	2 1/16	2 5/16

15° CUT FOR 30° TURN = WIDTH × .2679			45° CUT FOR 90° TURN = WIDTH × 1.000		
Size	One Piece "D"	Two Piece "E"	Size	One Piece "D"	Two Piece "E"
1/8" × 1"	1/4"	1/4"	1/8" × 1"	7/8"	1"
1/4 × 1 1/2	5/16	3/8	1/4 × 1 1/2	1 1/4	1 1/2
1/4 × 2	7/16	1/2	1/4 × 2	1 3/4	2
3/8 × 2 1/2	9/16	11/16	3/8 × 2 1/2	2 1/8	2 1/2
3/8 × 3	11/16	13/16	3/8 × 3	2 5/8	3
3/8 × 4	1	1 1/16	3/8 × 4	3 5/8	4

22½° CUT FOR 45° TURN = WIDTH × .414			67½° CUT FOR 135° TURN = WIDTH × 2.414		
Size	One Piece "D"	Two Piece "E"	Size	One Piece "D"	Two Piece "E"
1/8" × 1"	3/8"	7/16"	1/8" × 1"	2 1/8"	2 7/16"
1/4 × 1 1/2	1/2	5/8	1/4 × 1 1/2	3	3 5/8
1/4 × 2	3/4	13/16	1/4 × 2	4 1/4	4 13/16
3/8 × 2 1/2	7/8	1	3/8 × 2 1/2	5 1/8	6
3/8 × 3	1 1/16	1 1/4	3/8 × 3	6 5/16	7 1/4
3/8 × 4	1 1/2	1 5/8	3/8 × 4	8 3/4	9 5/8

SPECIAL OFFSETS (DRAWING #1)

Special offsets when the degree of rise & turn are known.

FORMULA: The cosine of degree of rise times the cosine of degree of turn equals the cosine of degree of elbow.

Find the degree of the bottom elbow:

The cosine of 45° rise is .707 times .866 the cosine of 30° turn equals .6123 the cosine of degree of elbow. From the trig tables the degree that has .6123 for its cosine is 52° – 14'. This is the degree of the bottom elbow.

The top elbow has a turn of 60° and is the complement of turn of the bottom elbow. The degree of rise always is the same for both elbows.

Find the degree of the top elbow:

The cosine of 45° rise is .707 times .500 the cosine of 60° turn equals .3535 the cosine of degree of elbow. From the trig tables the degree that has .3535 for its cosine is 69° – 18'. This is the degree of the top elbow.

Find the lengths of the sides of the 2 right triangles:

Use the 24" (SET) side of the 45° and figure the remaining sides. See pages 9 & 10 of this book under (ANGLE KNOWN) for method. Note that the (RUN) side of this angle is also the (TRAVEL) side of the 30° angle.

The (TRAVEL) side of the angle of rise is the true length of the offset center to center.

Find the cut length of pipe required:

Refer to pages 7 & 8 of this book and drawing #8 for method of calculating the end to center of above 2 elbows, as these must be subtracted from the center to center of offset to give you the cut length required.

ALL SIMILAR OFFSETS MAY BE CALCULATED USING THIS PROCEDURE.

SPECIAL OFFSETS
DRAWING # 1

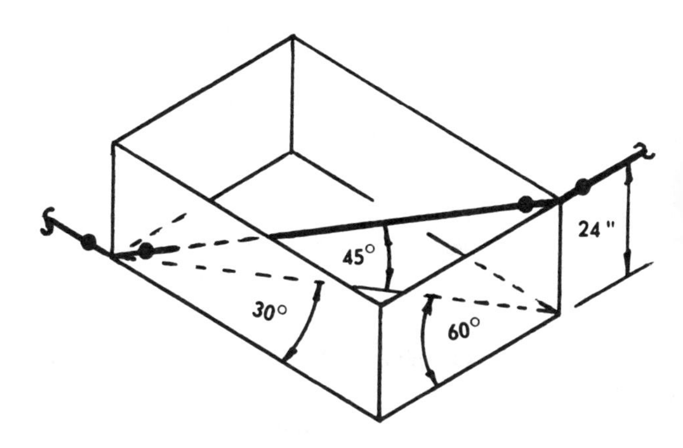

SPECIAL OFFSETS (DRAWING #2)

Special offsets when the degree of rise & turn are not known.

FORMULA: The cosine of degree of rise times the cosine of degree of turn equals the cosine of degree of elbow.

In this example you will have to use the dimensions of the 2 right triangles to figure the angles of rise and turn. Refer to pages 9 & 10 of this book under "TO FIND ANGLE". You will find that the angle of rise is 30° and the angle of turn is 22° - 30′. Use table (ANGLE KNOWN) for lengths of sides.

Using the cosine times cosine equals cosine formula:

The degree of the bottom elbow is 36° - 52′

The degree of the top elbow is 60°

Note that the top elbow is the complement of rise of the bottom elbow. 90°-30° = 60°. The degree of the top elbow. The (RUN) side of the 30° angle is also the (TRAVEL) side of the 22°-30′ angle.

The (TRAVEL) side of the angle of rise is the true length of the offset.

Find the cut length of pipe required:

Refer to pages 7& 8 of this book drawing #8 for method of calculating the centers of the above 2 elbows as these must be subtracted to give you the cut length required.

All similiar offsets may be calculated using this procedure.

Note that any 2 cosines used will call for the same degree of elbow regardless of their relationship.

SPECIAL OFFSETS
DRAWING # 2

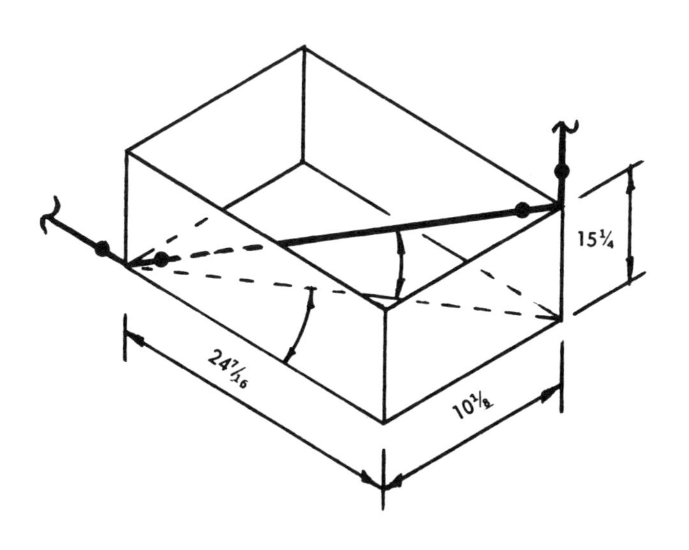

To simplify the fabrication, handling, and installation off all special type offsets, it is recommended that if at all possible lap joint flanges be installed at each end and in between the 2 elbows. In this way the fabricator can ignore the complex roll as well as having to match bolt holes.

On the opposite page are shown 6 various types of special offsets with the degree of elbow needed for the bottom and top.

The table below shows some standard angle combinations with their cosines multiplied to give the degree of elbow required. Note that the results will be the same regardless of which is the angle of rise and turn.

22½° / 22½° = 31°-24'	30° / 30° = 41°-24'	45° / 45° = 60°
22½° / 30° = 36°-52'	30° / 45° = 52°-14'	45° / 60° = 69°-18'
22½° / 45° = 49°-13'	30° / 60° = 64°-20'	45° / 67½° = 74°-18'
22½° / 60° = 62°-29'	30° / 67½° = 70°-39'	
22½° / 67½° = 69°-18'		

SPECIAL OFFSETS
SIX EXAMPLES

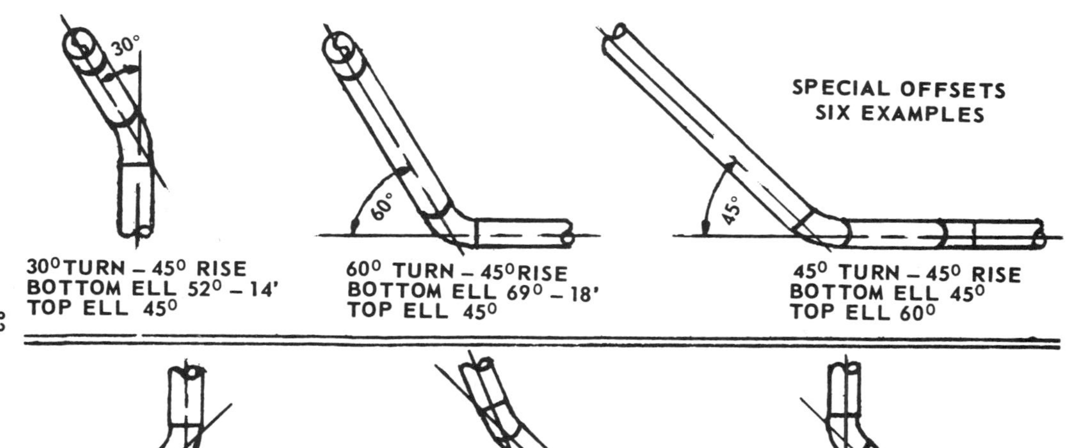

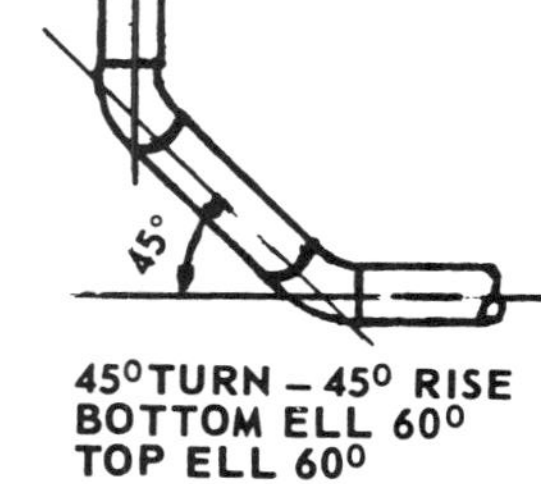
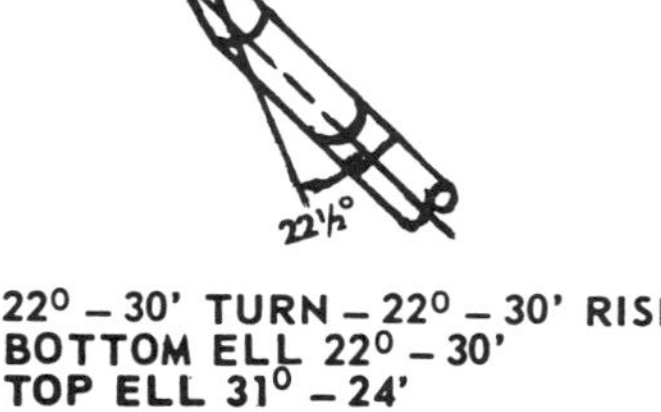
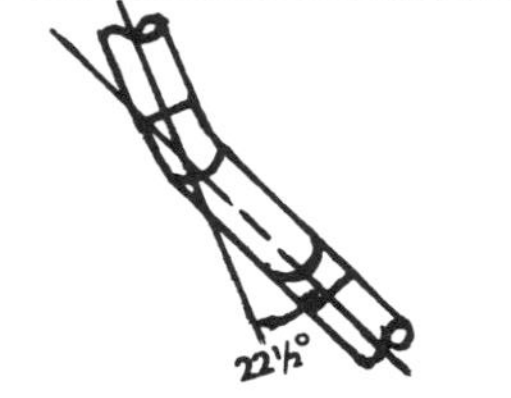

30° TURN – 45° RISE
BOTTOM ELL 52° – 14'
TOP ELL 45°

60° TURN – 45° RISE
BOTTOM ELL 69° – 18'
TOP ELL 45°

45° TURN – 45° RISE
BOTTOM ELL 45°
TOP ELL 60°

45° TURN – 45° RISE
BOTTOM ELL 45°
TOP ELL 60°

22° – 30' TURN – 22° – 30' RISE
BOTTOM ELL 22° – 30'
TOP ELL 31° – 24'

45° TURN – 45° RISE
BOTTOM ELL 60°
TOP ELL 60°

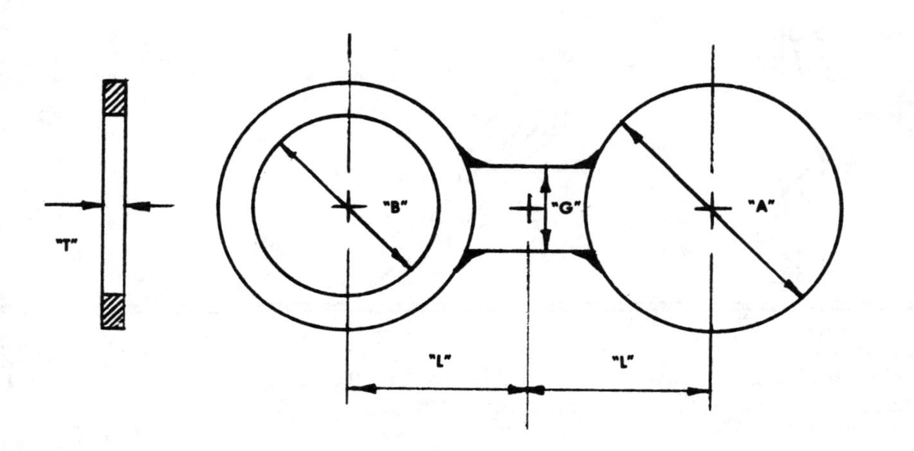

MATERIAL:

Carbon Steel, A-285-C or equal. Allow minimum of 1/8" for machining when ordering plate.

NOTES:

Thickness is based on formula (16) in ASA Code B31.3-1959.

SPECTACLE BLINDS 150 & 300# RF FLANGES CARBON STEEL PIPING

PIPE	150# RF – MAX. PRES. = 275 PSI @ 100° F					300# RF – MAX. PRES. = 720 PSI @ 100° F				
SIZE	A	B	T	L	G	A	B	T	L	G
1	2½	1 1/16	¼	1 9/16	1¼	2¾	1 1/16	¼	1¾	1½
1½	3¼	1⅝	¼	1 15/16	1½	3⅝	1⅝	¼	2¼	1½
2	4	2 3/32	¼	2⅜	1½	4¼	2 3/32	¼	2½	1
2½	4¾	2 15/32	¼	2¾	1½	5	2 15/32	¼	2 15/16	1½
3	5¼	3 3/32	¼	3	1½	5¾	3 3/32	⅜	3 5/16	1½
4	6¾	4 1/32	¼	3¾	1½	7	4 1/32	½	3 15/16	1½
6	8⅝	6 3/32	⅜	4¾	2	9¾	6 3/32	⅝	5 5/16	1¾
8	10⅞	8	½	5⅞	2	12	8	¾	6½	2
10	13¼	10 1/32	⅝	7⅛	2½	14⅛	10 1/32	1	7⅝	1½
12	16	12	¾	8½	2½	16½	12	1⅛	8⅞	2
14	17⅝	13¼	¾	9⅜	2¾	19	13¼	1¼	10⅛	1⅝
16	20⅛	15¼	⅞	10⅝	2¾	21⅛	15¼	1⅜	11¼	1¾
18	21½	17¼	1	11⅜	2¾	23⅜	17¼	1⅝	12⅜	1½
20	23¾	19¼	1⅛	12½	2¼	25⅝	19¼	1¾	13½	1¾
24	28⅛	23¼	1⅜	14¾	2¾	30⅜	23¼	2¼	16	2¼

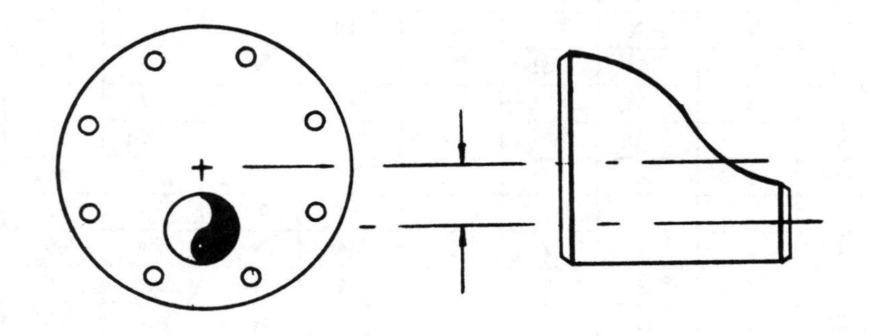

CENTER TO CENTER DIMENSIONS OF ECCENTRIC REDUCERS AND ECCENTRIC REDUCING FLANGES MADE FROM BLIND FLANGES.

CENTER TO CENTER DIMENSIONS OF ECCENTRIC B.W. REDUCERS AND DIMENSIONS FOR LAYING OUT CUTS FOR ECCENTRIC BLIND FLANGES

Size	Dimension	Size	Dimension	Size	Dimension
2 BY	1 = 1/2	6 BY	2 1/2 = 1 7/8	14 BY	6 = 3 11/16
	1 1/2 = 1/4		3 = 1 9/16		8 = 2 11/16
2 1/2 BY	1 = 3/4		4 = 1 1/16		10 = 1 5/8
	1 1/2 = 1/2	8 BY	3 = 2 9/16		12 = 5/8
	2 = 1/4		4 = 2 1/16	16 BY	6 = 4 11/16
3 BY	1 1/2 = 13/16		6 = 1		8 = 3 11/16
	2 = 9/16	10 BY	4 = 3 1/8		10 = 2 5/8
	2 1/2 = 5/16		6 = 2 1/16		12 = 1 5/8
4 BY	1 1/2 = 1 5/16		8 = 1 1/16		14 = 1
	2 = 1 1/16	12 BY	6 = 3 1/16	18 BY	8 = 4 11/16
	2 1/2 = 13/16		8 = 2 1/16		10 = 3 5/8
	3 = 1/2		10 = 1		12 = 2 5/8
					14 = 2
					16 = 1

DIMENSIONS FOR CUT OF 90° LONG RADIUS WELDELLS

FORMULA = RADIUS X DEGREES X .01745
IR = INSIDE RADIUS
OR = OUTSIDE RADIUS

OR
60°
IR
30°

RADIUS

7½°	=	"	x .1309
15°	=	"	x .2617
22½°	=	"	x .3926
30°	=	"	x .5235
45°	=	"	x .7852
60°	=	"	x 1.047

FORMULA = TANGENT OF ½ DEGREES OF TURN x RADIUS

15°	=	RADIUS	x .1316
22½°	=	"	x .1989
30°	=	"	x .2679
45°	=	"	x .414*
60°	=	"	x .5773

*REFER TO TABLES

E C

SIZE	22½°/67½°		30°/60°		45°/45°		END TO CENTER			
	IR	OR	IR	OR	IR	OR	15°	22½°	30°	60°
1½"	½	1¼	11/16	1 11/16	1	2½	5/16	7/16	5/8	1 5/16
2"	11/16	1⅝	15/16	2 3/16	1 7/16	3¼	⅜	⅝	13/16	1¾
3"	1 1/16	2 7/16	1 7/16	3¼	2 3/16	4⅞	9/16	⅞	1 3/16	2⅝
4"	1 7/16	3¼	1 15/16	4 5/16	2 15/16	6½	13/16	1 3/16	1⅝	3 7/16
6"	2¼	4 13/16	3	6 7/16	4½	9 11/16	1 3/16	1 13/16	2 7/16	5 3/16
8"	3	6⅜	4	8 9/16	6	12 13/16	1 9/16	2⅜	3 3/16	6 15/16
10"	3¾	8	5	10 11/16	7 9/16	16	2	3	4	8 11/16
12"	4 9/16	9 9/16	6 1/16	12¾	9⅛	19⅛	2⅜	3 9/16	4 13/16	10⅜
14"	5½	11	7 5/16	14 11/16	11	22	2¾	4 3/16	5⅝	12⅛
16"	6¼	12 9/16	8⅜	16¾	12 9/16	25⅛	3⅛	4¾	6 7/16	13⅞
18"	7 1/16	14 ⅛	9 7/16	18⅞	14⅛	28¼	3 9/16	5⅜	7¼	15 9/16
20"	7⅞	15 11/16	10 7/16	20 15/16	15 11/16	31 7/16	3 15/16	6	8	17 5/16
24"	9 7/16	18⅞	12 9/16	25⅛	18⅞	37 11/16	4¾	7 3/16	9⅝	20¾

LENGTH OF THREAD ON PIPE

LENGTH OF THREAD ON PIPE THAT IS SCREWED INTO VALVES OR FITTINGS TO MAKE A TIGHT JOINT

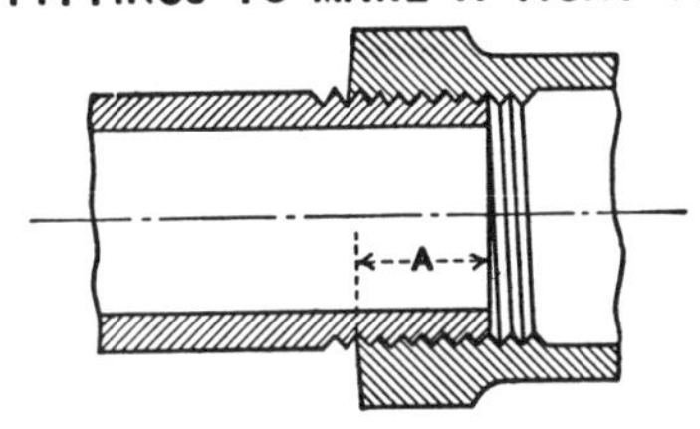

THREAD TAPER 1/16" PER INCH

PIPE SIZE	THREADS INCH	THREAD LENGTH	LENGTH "A"
1/8	27	7/16	5/16
1/4	18	5/8	7/16
3/8	18	5/8	7/16
1/2	14	13/16	9/16
3/4	14	13/16	9/16
1	11½	1	11/16
1¼	11½	1	11/16
1½	11½	1 1/32	11/16
2	11½	1 1/16	3/4
2½	8	1 9/16	1 1/16
3	8	1 5/8	1 1/8
4	8	1 3/4	1 3/16
6	8	1 15/16	1 3/8
8	8	2 3/16	1 7/16
10	8	2 3/8	1 5/8
12	8	2 9/16	1 3/4

WELDOLETS
SCHEDULE 40
MAKEUP – CENTER TO END

SIZE OF RUN

OUTLET SIZE	2	3	4	6	8	10	12
1	2 1/4	2 13/16	3 5/16	4 3/8	5 3/8	6 7/16	7 3/8
1 1/2	2 1/2	3 1/16	3 9/16	4 5/8	5 5/8	6 11/16	7 11/16
2	2 11/16	3 1/4	3 3/4	4 13/16	5 13/16	6 7/8	7 7/8
2 1/2		3 3/8	3 7/8	4 15/16	5 15/16	7	8
3		3 1/2	4	5 1/16	6 1/16	7 1/8	8 1/8
4			4 1/4	5 5/16	6 5/16	7 3/8	8 3/8
6				5 11/16	6 11/16	7 3/4	8 3/4
8					7 1/16	8 1/8	9 1/8
10						8 7/16	9 7/16

WELDOLETS
SCHEDULE 80

OUTLET SIZE	2	3	4	6	8	10	12
1	2 1/4	2 13/16	3 5/16	4 3/8	5 3/8	6 7/16	7 3/8
1 1/2	2 1/2	3 1/16	3 9/16	4 5/8	5 5/8	6 11/16	7 11/16
2	2 11/16	3 1/4	3 3/4	4 13/16	5 13/16	6 7/8	7 7/8
2 1/2		3 3/8	3 7/8	4 15/16	5 15/16	7	8
3		3 1/2	4	5 1/16	6 1/16	7 1/8	8 1/8
4			4 1/4	5 5/16	6 5/16	7 3/8	8 3/8
6				6 3/8	7 3/8	8 7/16	9 7/16
8					7 5/8	9 1/4	10 1/4
10						9 5/8	10 1/16

SOCKET WELD
ELLS, TEES, AND CROSSES
CENTER TO END AND LAYING LENGTHS

	CENTER TO END						LAYING LENGTH					
	½"	¾"	1"	1¼"	1½"	2"	½"	¾	1"	1¼"	1½"	2"
2000#	1⅛	1 5/16	1½	1¾	2	2⅜	⅝	¾	⅞	1 1/16	1¼	1½
3000#	1⅛	1 5/16	1½	1¾	2	2⅜	⅝	¾	⅞	1 1/16	1¼	1½
4000#	1 5/16	1½	1¾	2	2⅜	2½	¾	⅞	1 1/16	1¼	1½	1⅝
6000#	1 5/16	1½	1¾	2	2⅜	2½	¾	⅞	1 1/16	1¼	1½	1⅝

SCREWED
ELLS, TEES, AND CROSSES
CENTER TO END AND LAYING LENGTHS

	CENTER TO END						LAYING LENGTH					
	½"	¾"	1"	1¼"	1½"	2"	½"	¾"	1"	1¼"	1½"	2"
2000#	1⅛	1 5/16	1½	1¾	2	2⅜	9/16	¾	13/16	1 1/16	1 5/16	1⅝
3000#	1 5/16	1½	1¾	2	2⅜	2½	¾	15/16	1 1/16	1 5/16	1 11/16	1¾
6000#	1½	1¾	2	2⅜	2½	3¼	15/16	1 3/16	1 5/16	1 11/16	1 13/16	2½

FLANGED CAST STEEL VALVES
ASA FACE TO FACE DIMENSIONS, INCHES

CLASS	SIZE	TYPES OF VALVES						
		GATE	GLOBE	ANGLE ℄ TO FACE	CHECK	PLUG		BALL
						SHORT	REGULAR	
150 LB. 1/16 R.F.	1½	6½	6½	3¼	6½	6½		6½
	2	7	8	4	8	7		7
	2½	7½	8½	4¼	8½	7½		
	3	8	9½	4¾	9½	8		8
	3½	8½	10½		10½			
	4	9	11½	5¾	11½	9		9
	6	10½	16	8	14	10½	15½	10½
	8	11½	19½	9¾	19½	11½	18	11½
	10	13	24½	12¼	24½	13	21	13
	12	14	27½	13¾	27½	14	24	
300 LB. 1/16 R.F.	1½	7½	9	4½	9½	7½		7½
	2	8½	10½	5¼	10½	8½		8½
	2½	9½	11½	5¾	11½	9½		
	3	11⅛	12½	6¼	12½	11⅛		11⅛
	4	12	14	7	14	12		12
	5	15	15¾	7⅞	15¾			
	6	15⅞	17½	8¾	17½	15⅞	15⅞	15⅞
	8	16½	22	11	21	16½	19¾	16½
	10	18	24½	12¼	24½	18	22⅜	18
	12	19¾	28	14	28	19¾		
600 LB. ¼ R.F.	1½	9½	9½	4¾	9½		9½	
	2	11½	11½	5¾	11½		11½	11½
	2½	13	13	6½	13		13	
	3	14	14	7	14		14	14
	4	17	17	8½	17		17	17
	5	20	20	10	20			
	6	22	22	11	22		22	22
	8	26	26	13	26		26	26
	10	31	31	15½	31		31	31
	12	33	33	16½	33			33

CAST STEEL FLANGED FITTINGS
ELBOWS, TEES, & CROSSES

150 LB.		300 LB.	
SIZE	CENTER TO FACE	SIZE	CENTER TO FACE
1½	4	1½	4½
2	4½	2	5
2½	5	2½	5½
3	5½	3	6
3½	6	3½	6½
4	6½	4	7
5	7½	5	8
6	8	6	8½
8	9	8	10
10	11	10	11½
12	12	12	13
14	14	14	15
16	15	16	16½
18	16½	18	18
20	18	20	19½
24	22	24	22½

COMMERCIAL PIPE SIZES

NOMINAL PIPE SIZE	OUT-SIDE DIAM.	NOMINAL WALL					
		SCHED. 5S ①	SCHED. 10S ①	SCHED. 10	SCHED. 20	SCHED. 30	STAND-ARD ②
1/8	0.405	—	0.049	—	—	—	*0.068*
1/4	0.540	—	0.065	—	—	—	*0.088*
3/8	0.675	—	0.065	—	—	—	*0.091*
1/2	0.840	0.065	0.083	—	—	—	*0.109*
3/4	1.050	0.065	0.083	—	—	—	*0.113*
1	1.315	0.065	0.109	—	—	—	*0.133*
1 1/4	1.660	0.065	0.109	—	—	—	*0.140*
1 1/2	1.900	0.065	0.109	—	—	—	*0.145*
2	2.375	0.065	0.109	—	—	—	*0.154*
2 1/2	2.875	0.083	0.120	—	—	—	*0.203*
3	3.5	0.083	0.120	—	—	—	*0.216*
3 1/2	4.0	0.083	0.120	—	—	—	*0.226*
4	4.5	0.083	0.120	—	—	—	*0.237*
5	5.563	0.109	0.134	—	—	—	*0.258*
6	6.625	0.109	0.134	—	—	—	*0.280*
8	8.625	0.109	0.148	—	0.250	0.277	*0.322*
10	10.75	0.134	0.165	—	0.250	0.307	*0.365*
12	12.75	0.156	0.180	—	0.250	0.330	*0.375*
14 O.D.	14.0	0.156	0.188	0.250	0.312	0.375	0.375
16 O.D.	16.0	0.165	0.188	0.250	0.312	0.375	0.375
18 O.D.	18.0	0.165	0.188	0.250	0.312	0.438	0.375
20 O.D.	20.0	0.188	0.218	0.250	0.375	0.500	0.375
22 O.D.	22.0	0.188	0.218	0.250	0.375	0.500	0.375
24 O.D.	24.0	0.218	0.250	0.250	0.375	0.562	0.375
26 O.D.	26.0	—	—	0.312	0.500	—	0.375
28 O.D.	28.0	—	—	0.312	0.500	0.625	0.375
30 O.D.	30.0	0.250	0.312	0.312	0.500	0.625	0.375
32 O.D.	32.0	—	—	0.312	0.500	0.625	0.375
34 O.D.	34.0	—	—	0.312	0.500	0.625	0.375
36 O.D.	36.0	—	—	0.312	0.500	0.625	0.375
42 O.D.	42.0	—	—	—	—	—	0.375

NOTES:

① Schedules 5s and 10s are available in corrosion resistant materials and Schedule 10s is also available in carbon steel in sizes 12" and smaller.

② Thicknesses shown in italics are also available in stainless steel under the designation Schedule 40s.

AND WALL THICKNESSES ASA-B36.10 and B36.19

THICKNESS FOR								
SCHED. 40	SCHED. 60 ③	XS ④	SCHED. 80	SCHED. 100	SCHED. 120	SCHED. 140	SCHED. 160	XX STRONG
0.068	—	*0.095*	**0.095**	—	—	—	—	—
0.088	—	*0.119*	**0.119**	—	—	—	—	—
0.091	—	*0.126*	**0.126**	—	—	—	—	—
0.109	—	*0.147*	**0.147**	—	—	—	**0.188**	**0.294**
0.113	—	*0.154*	**0.154**	—	—	—	**0.219**	**0.308**
0.133	—	*0.179*	**0.179**	—	—	—	**0.250**	**0.358**
0.140	—	*0.191*	**0.191**	—	—	—	**0.250**	**0.382**
0.145	—	*0.200*	**0.200**	—	—	—	**0.281**	**0.400**
0.154	—	*0.218*	**0.218**	—	—	—	**0.344**	**0.436**
0.203	—	*0.276*	**0.276**	—	—	—	**0.375**	**0.552**
0.216	—	*0.300*	**0.300**	—	—	—	**0.438**	**0.600**
0.226	—	*0.318*	**0.318**	—	—	—	—	—
0.237	—	*0.337*	**0.337**	—	**0.438**	—	**0.531**	**0.674**
0.258	—	*0.375*	**0.375**	—	**0.500**	—	**0.625**	**0.750**
0.280	—	*0.432*	**0.432**	—	**0.562**	—	**0.719**	**0.864**
0.322	**0.406**	*0.500*	**0.500**	**0.594**	**0.719**	**0.812**	**0.906**	**0.875**
0.365	**0.500**	*0.500*	**0.594**	0.719	**0.844**	1.000	**1.125**	**1.000**
0.406	0.562	*0.500*	**0.688**	**0.844**	**1.000**	1.125	**1.312**	**1.000**
0.438	0.594	**0.500**	**0.750**	**0.938**	1.094	1.250	1.406	—
0.500	**0.656**	**0.500**	**0.844**	**1.031**	**1.219**	1.438	1.594	—
0.562	**0.750**	**0.500**	0.938	**1.156**	1.375	1.562	1.781	—
0.594	**0.812**	**0.500**	**1.031**	1.281	1.500	1.750	1.969	—
—	**0.875**	**0.500**	1.125	1.375	1.625	1.875	2.125	—
0.688	**0.969**	**0.500**	1.218	1.531	1.812	2.062	2.344	—
—	—	**0.500**	—	—	—	—	—	—
—	—	**0.500**	—	—	—	—	—	—
—	—	**0.500**	—	—	—	—	—	—
0.688	—	**0.500**	—	—	—	—	—	—
0.688	—	**0.500**	—	—	—	—	—	—
0.750	—	**0.500**	—	—	—	—	—	—
—	—	**0.500**	—	—	—	—	—	—

③ Thicknesses shown in light face for Schedule 60 and heavier pipe are not currently supplied by the mills, unless a certain minimum tonnage is ordered.

④ Thicknesses shown in italics are also available in stainless steel, under the designation Schedule 80s.

General Dimensions for

90° LONG RAD. WeldELL

90° REDUCING L.R. WeldELL

45° LONG RAD. WeldELL

180° LONG RADIUS WeldELL

Nom. Pipe Size	Outside Diam.	Nominal Wall Thickness				A	B
		STD ①	XS ②	160	XXS		
½	0.840	.109	.147	.187	.294	1½	⅝
¾	1.050	.113	.154	.218	.308	1⅛	7/16
1	1.315	.133	.179	.250	.358	1½	⅞
1¼	1.660	.140	.191	.250	.382	1⅞	1
1½	1.900	.145	.200	.281	.400	2¼	1⅛
2	2.375	.154	.218	.343	.436	3	1⅜
2½	2.875	.203	.276	.375	.552	3¾	1¾
3	3.500	.216	.300	.438	.600	4½	2
3½	4.000	.226	.318	......	.636 ●	5¼	2¼
4	4.500	.237	.337	.531	.674	6	2½
5	5.563	.258	.375	.625	.750	7½	3⅛
6	6.625	.280	.432	.718	.864	9	3¾
8	8.625	.322	.500	.906	.875	12	5
10	10.750	.365	.500	1.125	......	15	6¼
12	12.750	.375	.500	1.312	......	18	7½
14	14.000	.375	.500	1.406	......	21	8¾
16	16.000	.375	.500	1.593	......	24	10
18	18.000	.375	.500	1.781	......	27	11¼
20	20.000	.375	.500	1.968	......	30	12½
22	22.000	.375	.500	2.125	.....	33	13½
24	24.000	.375	.500	2.343	......	36	15
26 *	26.000	.375	.500	......	......	39	16
30 *	30.000	.375	.500	......	......	45	18½
36 *	36.000	.375	.500	......	......	54	22¼

① Standard wall thicknesses are the same as stainless steel schedule 40 s in sizes thru 12".

② Extra strong wall thicknesses are the same as stainless steel sched-ule 80 s in sizes thru 12".

Welding Fittings

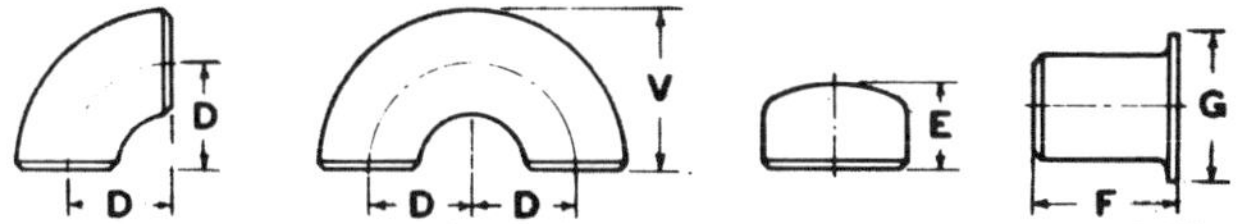

90° SHORT RAD. WeldELL — 180° SHORT RAD. WeldELL — CAP — LAP JOINT STUB END

K	D	V	E ③	E ④	G	F ASA	Nom. Pipe Size
1⅞	.....	.....	1	.....	1⅜	3	**½**
1¹¹⁄₁₆	.....	.....	1½	.....	1¹¹⁄₁₆	3	**¾**
2³⁄₁₆	1	1⅝	1½	1½	2	4	**1**
2¾	1¼	2¹⁄₁₆	1½	1½	2½	4	**1¼**
3¼	1½	2⁷⁄₁₆	1½	1½	2⅞	4	**1½**
4³⁄₁₆	2	3³⁄₁₆	1½	1¾	3⅝	6	**2**
5³⁄₁₆	2½	3¹⁵⁄₁₆	1½	2	4⅛	6	**2½**
6¼	3	4¾	2	2½	5	6	**3**
7¼	3½	5½	2½	3	5½	6	**3½**
8¼	4	6¼	2½	3	6³⁄₁₆	6	**4**
10⁵⁄₁₆	5	7¾	3	3½	7⁵⁄₁₆	8	**5**
12⁵⁄₁₆	6	9⁵⁄₁₆	3½	4	8½	8	**6**
16⁵⁄₁₆	8	12⁵⁄₁₆	4	5	10⅝	8	**8**
20⅜	10	15⅜	5	6	12¾	10	**10**
24⅜	12	18⅜	6	7	15	10	**12**
28	14	21	6½	7½	16¼	12	**14**
32	16	24	7	8	18½	12	**16**
36	18	27	8	9	21	12	**18**
40	20	30	9	10	23	12	**20**
44	.....	.....	10	10	25¼	12	**22**
48	24	36	10½	12	27¼	12	**24**
52	.....	.....	10½	.....	.....	.....	**26** *
60	30	45	10½	.....	.....	.....	**30** *
.....	36	54	10½	.....	.....	.....	**36** *

③ Applies for XS wall thickness and less.

④ Applies for wall thickness greater than XS.

* This size not covered by ASA B16.9.

• This size not covered by ASA B36.10.

General Dimensions for

STRAIGHT TEE

M

C C

REDUCING TEE

M

C C

Nom. Pipe Size	Outlet	Outside Diam.	Nominal Wall Thickness				C	M	H
			STD	XS	160	XXS			
½	½	.840	.109	.147	.187	.294	1	1	...
	⅜	.675	.091	.126	....	....	1	1	...
¾	¾	1.050	.113	.154	.218	.308	1⅛	1⅛	...
	½	.840	.109	.147	.187	.294	1⅛	1⅛	1½
1	1	1.315	.133	.179	.250	.358	1½	1½	...
	¾	1.050	.113	.154	.218	.308	1½	1½	2
	½	.840	.109	.147	.187	.294	1½	1½	2
1¼	1¼	1.660	.140	.191	.250	.382	1⅞	1⅞	...
	1	1.315	.133	.179	.250	.358	1⅞	1⅞	2
	¾	1.050	.113	.154	.218	.308	1⅞	1⅞	2
	½	.840	.109	.147	.187	.294	1⅞	1⅞	2
1½	1½	1.900	.145	.200	.281	.400	2¼	2¼	...
	1¼	1.660	.140	.191	.250	.382	2¼	2¼	2½
	1	1.315	.133	.179	.250	.358	2¼	2¼	2½
	¾	1.050	.113	.154	.218	.308	2¼	2¼	2½
	½	.840	.109	.147	.187	.294	2¼	2¼	2½
2	2	2.375	.154	.218	.343	.436	2½	2½	....
	1½	1.900	.145	.200	.281	.400	2½	2⅜	3
	1¼	1.660	.140	.191	.250	.382	2½	2¼	3
	1	1.315	.133	.179	.250	.358	2½	2	3
	¾	1.050	.113	.154	.218	.308	2½	1¾	3
2½	2½	2.875	.203	.276	.375	.552	3	3	...
	2	2.375	.154	.218	.343	.436	3	2¾	3½
	1½	1.900	.145	.200	.281	.400	3	2⅝	3½
	1¼	1.660	.140	.191	.250	.382	3	2½	3½
	1	1.315	.133	.179	.250	.358	3	2¼	3½

†This size not covered by ASA B36.10

ASA B16.9

ASA B36.10

Welding Fittings

CONCENTRIC REDUCER

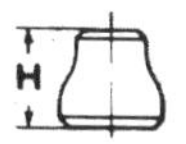

ECCENTRIC REDUCER

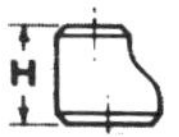

Nom. Pipe Size	Outlet	Outside Diam.	Nominal Wall Thickness				C	M	H
			STD	XS	160	XXS			
3	3	3.500	.216	.300	.438	.600	3⅜	3⅜	...
	2½	2.875	.203	.276	.375	.552	3⅜	3¼	3½
	2	2.375	.154	.218	.343	.436	3⅜	3	3½
	1½	1.900	.145	.200	.281	.400	3⅜	2⅞	3½
	1¼	1.660	.140	.191	.250	.382	3⅜	2¾	3½
3½	3½	4.000	.226	.318	....	.636 †	3¾	3¾	...
	3	3.500	.216	.300	.438	.600	3¾	3⅝	4
	2½	2.875	.203	.276	.375	.552	3¾	3½	4
	2	2.375	.154	.218	.343	.436	3¾	3¼	4
	1½	1.900	.145	.200	.281	.400	3¾	3⅛	4
4	4	4.500	.237	.337	.531	.674	4⅛	4⅛	...
	3½	4.000	.226	.318	....	.636 †	4⅛	4	4
	3	3.500	.216	.300	.438	.600	4⅛	3⅞	4
	2½	2.875	.203	.276	.375	.552	4⅛	3¾	4
	2	2.375	.154	.218	.343	.436	4⅛	3½	4
	1½	1.900	.145	.200	.281	.400	4⅛	3⅜	4
5	5	5.563	.258	.375	.625	.750	4⅞	4⅞	...
	4	4.500	.237	.337	.531	.674	4⅞	4⅝	5
	3½	4.000	.226	.318	....	.636 †	4⅞	4½	5
	3	3.500	.216	.300	.438	.600	4⅞	4⅜	5
	2½	2.875	.203	.276	.375	.552	4⅞	4¼	5
	2	2.375	.154	.218	.343	.436	4⅞	4⅛	5
6	6	6.625	.280	.432	.718	.864	5⅝	5⅝	...
	5	5.563	.258	.375	.625	.750	5⅝	5⅜	5½
	4	4.500	.237	.337	.531	.674	5⅝	5⅛	5½
	3½	4.000	.226	.318	....	.636 †	5⅝	5	5½
	3	3.500	.216	.300	.438	.600	5⅝	4⅞	5½
	2½	2.875	.203	.276	.375	.552	5⅝	4¾	5½

General Dimensions for

STRAIGHT TEE — C, C, M

REDUCING TEE — C, C, M

Nom. Pipe Size	Outlet	Outside Diam.	Nominal wall Thickness				C	M	H
			STD	XS	160	XXS			
8	8	8.625	.322	.500	.906	.875	7	7	...
	6	6.625	.280	.432	.718	.864	7	6⅝	6
	5	5.563	.258	.375	.625	.750	7	6⅜	6
	4	4.500	.237	.337	.531	.674	7	6⅛	6
	3½	4.000	.226	.318	....	.636 †	7	6	6
10	10	10.750	.365	.500	1.125	....	8½	8½	...
	8	8.625	.322	.500	.906	....	8½	8	7
	6	6.625	.280	.432	.718	....	8½	7⅝	7
	5	5.563	.258	.375	.625	....	8½	7½	7
	4	4.500	.237	.337	.531	....	8½	7¼	7
12	12	12.750	.375	.500	1.312	....	10	10	...
	10	10.750	.365	.500	1.125	....	10	9½	8
	8	8.625	.322	.500	.906	....	10	9	8
	6	6.625	.280	.432	.718	....	10	8⅝	8
	5	5.563	.258	.375	.625	....	10	8½	8
14	14	14.000	.375	.500	1.406	....	11	11	...
	12	12.750	.375	.500	1.312	....	11	10⅝	13
	10	10.750	.365	.500	1.125	....	11	10⅛	13
	8	8.625	.322	.500	.906	....	11	9¾	13
	6	6.625	.280	.432	.718	....	11	9⅜	13
16	16	16.000	.375	.500	1.593	....	12	12	...
	14	14.000	.375	.500	1.406	....	12	12	14
	12	12.750	.375	.500	1.312	....	12	11⅝	14
	10	10.750	.365	.500	1.125	....	12	11⅛	14
	8	8.625	.322	.500	.906	....	12	10¾	14
	6	6.625	.280	.432	.718	....	12	10⅜	14
18	18	18.000	.375	.500	1.781	....	13½	13½	...
	16	16.000	.375	.500	1.593	....	13½	13	15
	14	14.000	.375	.500	1.406	....	13½	13	15

ASA B16.9

ASA B36.10

Welding Fittings

CONCENTRIC REDUCER

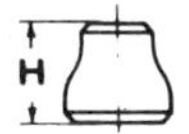

ECCENTRIC REDUCER

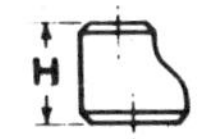

Nom. Pipe Size	Outlet	Outside Diam.	Nominal Wall Thickness				C	M	H
			STD	XS	160	XXS			
18	12	12.750	.375	.500	1.312	....	13½	12⅝	15
	10	10.750	.365	.500	1.125	....	13½	12⅛	15
	8	8.625	.322	.500	.906	....	13½	11¾	15
20	20	20.000	.375	.500	1.968	....	15	15	...
	18	18.000	.375	.500	1.781	....	15	14½	20
	16	16.000	.375	.500	1.593	....	15	14	20
	14	14.000	.375	.500	1.406	....	15	14	20
	12	12.750	.375	.500	1.312	....	15	13⅝	20
	10	10.750	.365	.500	1.125	....	15	13⅛	20
	8	8.625	.322	.500	.906	....	15	12¾	20
22	22	22.000	.375	.500	2.125	....	16½	16½	...
	20	20.000	.375	.500	1.968	....	16½	16	20
	18	18.000	.375	.500	1.781	....	16½	15½	20
	16	16.000	.375	.500	1.593	....	16½	15	20
	14	14.000	.375	.500	1.406	....	16½	15	20
	12	12.750	.375	.500	1.312	....	16½	14⅝	...
	10	10.750	.365	.500	1.125	....	16½	14⅛	...
24	24	24.000	.375	.500	2.343	....	17	17	...
	22	22.000	.375	.500	2.125	....	17	17	20
	20	20.000	.375	.500	1.968	....	17	17	20
	18	18.000	.375	.500	1.781	....	17	16½	20
	16	16.000	.375	.500	1.593	....	17	16	20
	14	14.000	.375	.500	1.406	....	17	16	20
	12	12.750	.375	.500	1.312	....	17	15⅝	20
	10	10.750	.365	.500	1.125	....	17	15⅛	20
30*	30	30.000	.375	.500	....	....	22	22	...
	24	24.000	.375	.500	2.343	....	22	21	24
	22	22.000	.375	.500	2.125	....	22	20½	24
	20	20.000	.375	.500	1.968	....	22	20	24
	18	18.000	.375	.500	1.781	....	22	19½	...
	16	16.000	.375	.500	1.593	....	22	19	...

*This size not covered by ASA B16.9

†This size not covered by ASA B36.10

General Dimensions for

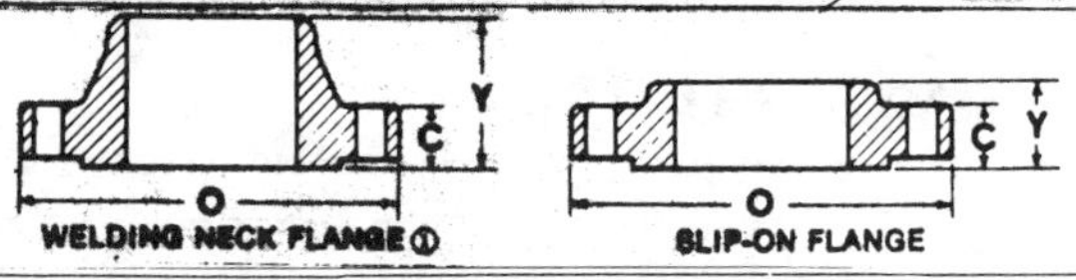

125 lb. LW　　ASA B16.1　　ASA B16.5　　A-181-I

Nom. Pipe Size	Flange O.D. O	Flange Thickness C ②	Diam. of Raised Face	Length Thru Hub Y: Welding Neck ①	Length Thru Hub Y: Slip-on	Drilling: No. & Size of Holes	Drilling: Bolt Circle	Bores: Welding Neck ①	Bores: Slip-on
½	...	....	....	....	....	....	....	....	....
¾	...	....	....	....	....	....	....	....	....
1	4¼	⅜	2	1¼	....	4- ⅝	3⅛	1.05	....
1¼	4⅝	⅜	2⅜	1¼	....	4- ⅝	3½	1.38	....
1½	5	⅜	2⅝	1¼	....	4- ⅝	3⅞	1.61	....
2	6	7/16	3	1⅜	....	4- ¾	4¾	2.07	....
2½	7	7/16	3½	1⅜	⅞	4- ¾	5½	2.47	2.94
3	7½	½	4	1⅝	⅞	4- ¾	6	3.07	3.57
3½	8½	½	4¾	1⅝	....	8- ¾	7	3.55	....
4	9	½	5½	1⅝	⅞	8- ¾	7½	4.13	4.57
5	10	9/16	6½	1⅝	⅞	8- ⅞	8½	5.05	5.66
6	11	9/16	7½	1⅞	1¼	8- ⅞	9½	6.19	6.72
8	13½	9/16	9½	1⅞	1¼	8- ⅞	11¾	8.19	8.72
10	16	11/16	11¾	2⅛	1¼	12-1	14¼	10.31	10.88
12	19	11/16	13¾	2¼	1¼	12-1	17	12.25	12.88
14	21	¾	....	....	1¼	12-1⅛	18¾	....	14.14
16	23½	¾	....	....	1¼	16-1⅛	21¼	....	16.16
18	25	¾	....	....	1¼	16-1¼	22¾	....	18.18
20	27½	¾	....	....	1¼	20-1¼	25	....	20.20
24	32	1	....	....	1¾	20-1⅜	29½	....	24.25

① Welding neck flange sizes 5", 3½" and smaller are bored for standard weight pipe. Sizes 4", 6" and larger are bored to match light wall pipe and gas distribution welding fittings. Slip-on flanges are bored to match O.D. of light wall pipe and gas distribution welding fittings.

② All sizes are regularly furnished with flat face. A 1/32" raised face can be furnished on request.

Forged Steel Flanges

THREADED FLANGE — LAP JOINT FLANGE (C, Y, O)

ASA B16.5 **A-181-I** **150 lb.**

Nom. Pipe Size	Flange O.D. O	Flange Thickness C ②	Diam. of Raised Face	Length Thru Hub Y ②			Drilling		Bores	
				Welding Neck ①	Slip-on Thread. and Socket ①	Lap Joint	No. & Size of Holes	Bolt Circle	Slip-on	Lap Joint
½	3½	7/16	1⅜	1⅞	⅝	⅝	4- ⅝	2⅜	.88	.90
¾	3⅞	½	1 11/16	2 1/16	⅝	⅝	4- ⅝	2¾	1.09	1.11
1	4¼	9/16	2	2 3/16	11/16	11/16	4- ⅝	3⅛	1.36	1.38
1¼	4⅝	⅝	2½	2¼	13/16	13/16	4- ⅝	3½	1.70	1.72
1½	5	11/16	2⅞	2 7/16	⅞	⅞	4- ⅝	3⅞	1.95	1.97
2	6	¾	3⅝	2½	1	1	4- ¾	4¾	2.44	2.46
2½	7	⅞	4⅛	2¾	1⅛	1⅛	4- ¾	5½	2.94	2.97
3	7½	15/16	5	2¾	1 3/16	1 3/16	4- ¾	6	3.57	3.60
3½	8½	15/16	5½	2 13/16	1¼	1¼	8- ¾	7	4.07	4.10
4	9	15/16	6 3/16	3	1 5/16	1 5/16	8- ¾	7½	4.57	4.60
5	10	15/16	7 5/16	3½	1 7/16	1 7/16	8- ⅞	8½	5.66	5.69
6	11	1	8½	3½	1 9/16	1 9/16	8- ⅞	9½	6.72	6.75
8	13½	1⅛	10⅝	4	1¾	1¾	8- ⅞	11¾	8.72	8.75
10	16	1 3/16	12¾	4	1 15/16	1 15/16	12-1	14¼	10.88	10.92
12	19	1¼	15	4½	2 3/16	2 3/16	12-1	17	12.88	12.92
14	21	1⅜	16¼	5	2¼	3⅛	12-1⅛	18¾	14.14	14.18
16	23½	1 7/16	18½	5	2½	3 7/16	16-1⅛	21¼	16.16	16.19
18	25	1 9/16	21	5½	2 11/16	3 13/16	16-1¼	22¾	18.18	18.20
20	27½	1 11/16	23	5 11/16	2⅞	4 1/16	20-1¼	25	20.20	20.25
24	32	1⅞	27¼	6	3¼	4⅜	20-1⅜	29½	24.25	24.25

NOTES:

① Always specify bore when ordering.

② Includes 1/16" raised face in 150# & 300# standards. Does NOT include ¼" raised face in 400# and heavier standards.

General Dimensions for

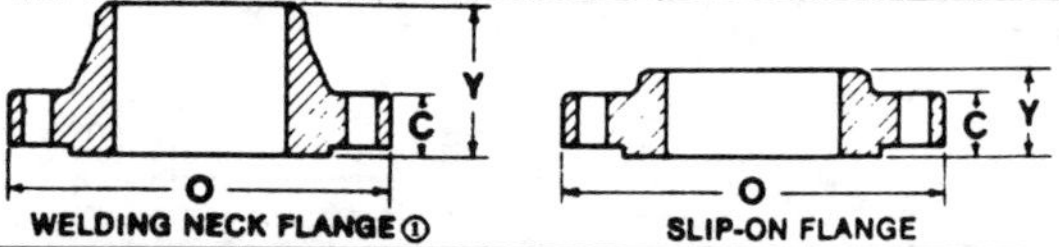

300 lb. ASA B16.5 A-181-I

Nom. Pipe Size	Flange O.D. O	Flange Thickness C ②	Diam. of Raised Face	Length Thru Hub Y ② Welding Neck ①	Slip-on Thread. and Socket ①	Lap Joint	Drilling No. & Size of Holes	Bolt Circle	Bores Slip-on	Lap Joint
½	3¾	9/16	1⅜	2 1/16	⅞	⅞	4- ⅝	2⅝	.88	.90
¾	4⅝	⅝	1 11/16	2¼	1	1	4- ¾	3¼	1.09	1.11
1	4⅞	11/16	2	2 7/16	1 1/16	1 1/16	4- ¾	3½	1.36	1.38
1¼	5¼	¾	2½	2 9/16	1 1/16	1 1/16	4- ¾	3⅞	1.70	1.72
1½	6⅛	13/16	2⅞	2 11/16	1 3/16	1 3/16	4- ⅞	4½	1.95	1.97
2	6½	⅞	3⅝	2¾	1 5/16	1 5/16	8- ¾	5	2.44	2.46
2½	7½	1	4⅛	3	1½	1½	8- ⅞	5⅞	2.94	2.97
3	8¼	1⅛	5	3⅛	1 11/16	1 11/16	8- ⅞	6⅝	3.57	3.60
3½	9	1 3/16	5½	3 3/16	1¾	1¾	8- ⅞	7¼	4.07	4.10
4	10	1¼	6 3/16	3⅜	1⅞	1⅞	8- ⅞	7⅞	4.57	4.60
5	11	1⅜	7 5/16	3⅞	2	2	8- ⅞	9¼	5.66	5.69
6	12½	1 7/16	8½	3⅞	2 1/16	2 1/16	12- ⅞	10⅝	6.72	6.75
8	15	1⅝	10⅝	4⅜	2 7/16	2 7/16	12-1	13	8.72	8.75
10	17½	1⅞	12¾	4⅝	2⅝	3¾	16-1⅛	15¼	10.88	10.92
12	20½	2	15	5⅛	2⅞	4	16-1¼	17¾	12.88	12.92
14	23	2⅛	16¼	5⅝	3	4⅜	20-1¼	20¼	14.14	14.18
16	25½	2¼	18½	5¾	3¼	4¾	20-1⅜	22½	16.16	16.19
18	28	2⅜	21	6¼	3½	5⅛	24-1⅜	24¾	18.18	18.20
20	30½	2½	23	6⅜	3¾	5½	24-1⅜	27	20.20	20.25
24	36	2¾	27¼	6⅝	4 3/16	6	24-1⅝	32	24.25	24.25

Forged Steel Flanges

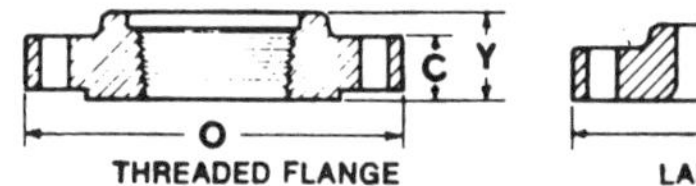

THREADED FLANGE

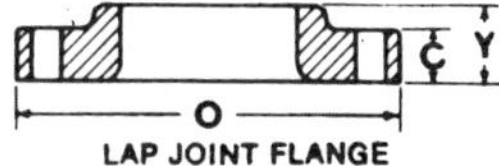

LAP JOINT FLANGE

A-105-I **400 lb.**

Nom. Pipe Size	Flange O.D. O	Flange Thickness C ②	Diam. of Raised Face	Length Thru Hub Y ②			Drilling		Bores	
				Welding Neck ①	Slip-on and Thread.	Lap Joint	No. & Size of Holes	Bolt Circle	Slip-on	Lap Joint
½	3¾	9/16	1⅜	2 1/16	⅞	⅞	4- ⅝	2⅝	.88	.90
¾	4⅝	⅝	1 11/16	2¼	1	1	4- ¾	3¼	1.09	1.11
1	4⅞	11/16	2	2 7/16	1 1/16	1 1/16	4- ¾	3½	1.36	1.38
1¼	5¼	13/16	2½	2⅝	1⅛	1⅛	4- ¾	3⅞	1.70	1.72
1½	6⅛	⅞	2⅞	2¾	1¼	1¼	4- ⅞	4½	1.95	1.97
2	6½	1	3⅝	2⅞	1 7/16	1 7/16	8- ¾	5	2.44	2.46
2½	7½	1⅛	4⅛	3⅛	1⅝	1⅝	8- ⅞	5⅞	2.94	2.97
3	8¼	1¼	5	3¼	1 13/16	1 13/16	8- ⅞	6⅝	3.57	3.60
3½	9	1⅜	5½	3⅜	1 15/16	1 15/16	8-1	7¼	4.07	4.10
4	10	1⅜	6 3/16	3½	2	2	8-1	7⅞	4.57	4.60
5	11	1½	7 5/16	4	2⅛	2⅛	8-1	9¼	5.66	5.69
6	12½	1⅝	8½	4 1/16	2¼	2¼	12-1	10⅝	6.72	6.75
8	15	1⅞	10⅝	4⅝	2 11/16	2 11/16	12-1⅛	13	8.72	8.75
10	17½	2⅛	12¾	4⅞	2⅞	4	16-1¼	15¼	10.88	10.92
12	20½	2¼	15	5⅜	3⅛	4¼	16-1⅜	17¾	12.88	12.92
14	23	2⅜	16¼	5⅞	3 5/16	4⅝	20-1⅜	20¼	14.14	14.18
16	25½	2½	18½	6	3 11/16	5	20-1½	22½	16.16	16.19
18	28	2⅝	21	6½	3⅞	5⅜	24-1½	24¾	18.18	18.20
20	30½	2¾	23	6⅝	4	5¾	24-1⅝	27	20.20	20.25
24	36	3	27¼	6⅞	4½	6¼	24-1⅞	32	24.25	24.25

NOTES:

① Always specify bore when ordering.

② Includes 1/16" raised face in 150# & 300# standards. Does NOT include ¼" raised face in 400# and heavier standards.

General Dimensions for

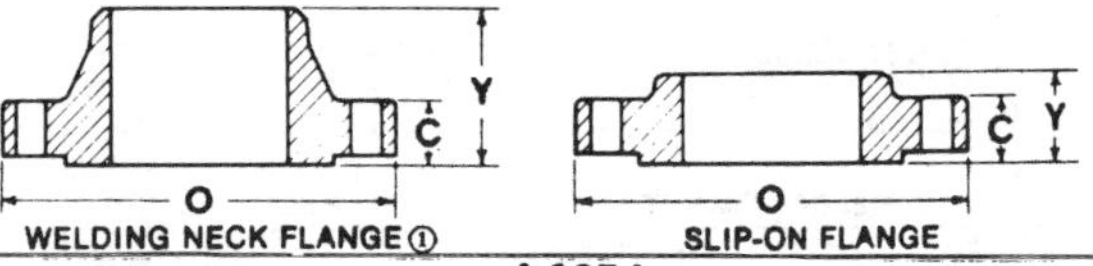

600 lb. **A-105-I**

Nom. Pipe Size	Flange O.D. O	Flange Thickness C ②	Diam. of Raised Face	Length Thru Hub Y ②			Drilling		Bores	
				Welding Neck ①	Slip-on Thread. and Socket ①	Lap Joint	No. & Size of Holes	Bolt Circle	Slip-on	Lap Joint
½	3¾	9/16	1⅜	2 1/16	⅞	⅞	4- ⅝	2⅝	.88	.90
¾	4⅝	⅝	1 11/16	2¼	1	1	4- ¾	3¼	1.09	1.11
1	4⅞	11/16	2	2 7/16	1 1/16	1 1/16	4- ¾	3½	1.36	1.38
1¼	5¼	13/16	2½	2⅝	1⅛	1⅛	4- ¾	3⅞	1.70	1.72
1½	6⅛	⅞	2⅞	2¾	1¼	1¼	4- ⅞	4½	1.95	1.97
2	6½	1	3⅝	2⅞	1 7/16	1 7/16	8- ¾	5	2.44	2.46
2½	7½	1⅛	4⅛	3⅛	1⅝	1⅝	8- ⅞	5⅞	2.94	2.97
3	8¼	1¼	5	3¼	1 13/16	1 13/16	8- ⅞	6⅝	3.57	3.60
3½	9	1⅜	5½	3⅜	1 15/16	1 15/16	8-1	7¼	4.07	4.10
4	10¾	1½	6 3/16	4	2⅛	2⅛	8-1	8½	4.57	4.60
5	13	1¾	7 5/16	4½	2⅜	2⅜	8-1⅛	10½	5.66	5.69
6	14	1⅞	8½	4⅝	2⅝	2⅝	12-1⅛	11½	6.72	6.75
8	16½	2 3/16	10⅝	5¼	3	3	12-1¼	13¾	8.72	8.75
10	20	2½	12¾	6	3⅜	4⅜	16-1⅜	17	10.88	10.92
12	22	2⅝	15	6⅛	3⅝	4⅝	20-1⅜	19¼	12.88	12.92
14	23¾	2¾	16¼	6½	3 11/16	5	20-1½	20¾	14.14	14.18
16	27	3	18½	7	4 3/16	5½	20-1⅝	23¾	16.16	16.19
18	29¼	3¼	21	7¼	4⅝	6	20-1¾	25¾	18.18	18.20
20	32	3½	23	7½	5	6½	24-1¾	28½	20.20	20.25
24	37	4	27¼	8	5½	7¼	24-2	33	24.25	24.25

Forged Steel Flanges

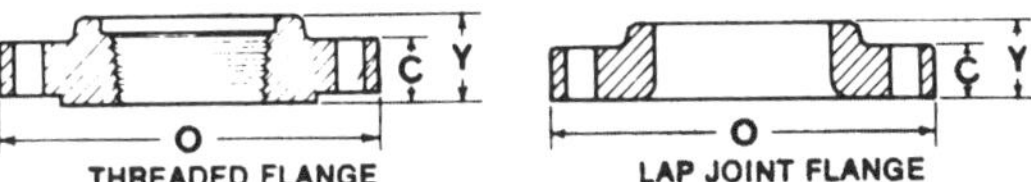

ASA B16.5 **A-105-II** **900 lb.**

Nom. Pipe Size	Flange O.D. O	Flange Thick-ness C ②	Diam. of Raised Face	Length Thru Hub Y ②			Drilling		Bores	
				Weld-ing Neck ①	Slip-on and Thread.	Lap Joint	No. & Size of Holes	Bolt Circle	Slip-on	Lap Joint
½	4¾	⅞	1⅜	2⅜	1¼	1¼	4- ⅞	3¼	.88	.90
¾	5⅛	1	1¹¹⁄₁₆	2¾	1⅜	1⅜	4- ⅞	3½	1.09	1.11
1	5⅞	1⅛	2	2⅞	1⅝	1⅝	4-1	4	1.36	1.38
1¼	6¼	1⅛	2½	2⅞	1⅝	1⅝	4-1	4⅜	1.70	1.72
1½	7	1¼	2⅞	3¼	1¾	1¾	4-1⅛	4⅞	1.95	1.97
2	8½	1½	3⅝	4	2¼	2¼	8-1	6½	2.44	2.46
2½	9⅝	1⅝	4⅛	4⅛	2½	2½	8-1⅛	7½	2.94	2.97
3	9½	1½	5	4	2⅛	2⅛	8-1	7½	3.57	3.60
3½	...	...	...	...	...	...	....	...	...	...
4	11½	1¾	6³⁄₁₆	4½	2¾	2¾	8-1¼	9¼	4.57	4.60
5	13¾	2	7⁵⁄₁₆	5	3⅛	3⅛	8-1⅜	11	5.66	5.69
6	15	2³⁄₁₆	8½	5½	3⅜	3⅜	12-1¼	12½	6.72	6.75
8	18½	2½	10⅝	6⅜	4	4½	12-1½	15½	8.72	8.75
10	21½	2¾	12¾	7¼	4¼	5	16-1½	18½	10.88	10.92
12	24	3⅛	15	7⅞	4⅝	5⅝	20-1½	21	12.88	12.92
14	25¼	3⅜	16¼	8⅜	5⅛	6⅛	20-1⅝	22	14.14	14.18
16	27¾	3½	18½	8½	5¼	6½	20-1¾	24¼	16.16	16.19
18	31	4	21	9	6	7½	20-2	27	18.18	18.20
20	33¾	4¼	23	9¾	6¼	8¼	20-2⅛	29½	20.20	20.25
24	41	5½	27¼	11½	8	10½	20-2⅝	35½	24.25	24.25

NOTES:

① Always specify bore when ordering.

② Includes ¹⁄₁₆" raised face in 150# & 300# standards. Does NOT include ¼" raised face in 400# and heavier standards.

General Dimensions for

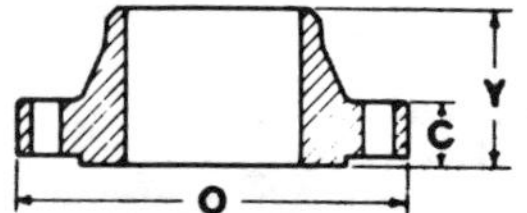

WELDING NECK FLANGE ①

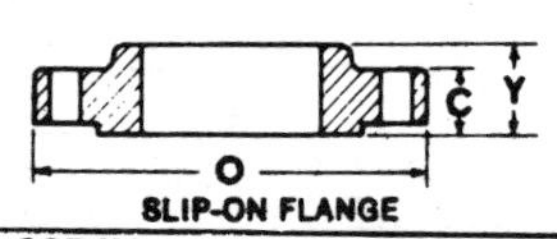

SLIP-ON FLANGE

1500 lb. **A-105-II**

Nom. Pipe Size	Flange O.D. O	Flange Thickness C ②	Diam. of Raised Face	Length Thru Hub Y ②			Drilling		Bores	
				Welding Neck ①	Slip-on Thread. and Socket ①	Lap Joint	No. & Size of Holes	Bolt Circle	Slip-on	Lap Joint
½	4¾	⅞	1⅜	2⅜	1¼	1¼	4- ⅞	3¼	.88	.90
¾	5⅛	1	1 11/16	2¾	1⅜	1⅜	4- ⅞	3½	1.09	1.11
1	5⅞	1⅛	2	2⅞	1⅝	1⅝	4-1	4	1.36	1.38
1¼	6¼	1⅛	2½	2⅞	1⅝	1⅝	4-1	4⅜	1.70	1.72
1½	7	1¼	2⅞	3¼	1¾	1¾	4-1⅛	4⅞	1.95	1.97
2	8½	1½	3⅝	4	2¼	2¼	8-1	6½	2.44	2.46
2½	9⅝	1⅝	4⅛	4⅛	2½	2½	8-1⅛	7½	2.94	2.97
3	10½	1⅞	5	4⅝	2⅞	2⅞	8-1¼	8	3.57	3.60
3½	...	...	...	...	...	...	....	...	...	...
4	12¼	2⅛	6 3/16	4⅞	3 9/16	3 9/16	8-1⅜	9½	4.57	4.60
5	14¾	2⅞	7 5/16	6⅛	4⅛	4⅛	8-1⅝	11½	5.66	5.69
6	15½	3¼	8½	6¾	4 11/16	4 11/16	12-1½	12½	6.72	6.75
8	19	3⅝	10⅝	8⅜	5⅝	5⅝	12-1¾	15½	8.72	8.75
10	23	4¼	12¾	10	6¼	7	12-2	19	10.88	10.92
12	26½	4⅞	15	11⅛	7⅛	8⅝	16-2⅛	22½	12.88	12.92
14	29½	5¼	16¼	11¾	...	9½	16-2⅜	25	14.14	14.18
16	32½	5¾	18½	12¼	...	10¼	16-2⅝	27¾	16.16	16.19
18	36	6⅜	21	12⅞	...	10⅞	16-2⅞	30½	18.18	18.20
20	38¾	7	23	14	...	11½	16-3⅛	32¾	20.20	20.25
24	46	8	27¼	16	...	13	16-3⅝	39	24.25	24.25

Forged Steel Flanges

THREADED FLANGE

LAP JOINT FLANGE

ASA B16.5 **A-105-II** **2500 lb.**

Nom. Pipe Size	Flange O.D. O	Flange Thickness C (2)	Diam. of Raised Face	Length Thru Hub Y (2)			Drilling		Bores	
				Welding Neck (1)	Slip-on and Thread.	Lap Joint	No. & Size of Holes	Bolt Circle	Slip-on	Lap Joint
½	5¼	1 3/16	1⅜	2⅞	1 9/16	1 9/16	4-⅞	3½	.88	.90
¾	5½	1¼	1 11/16	3⅛	1 11/16	1 11/16	4-⅞	3¾	1.09	1.11
1	6¼	1⅜	2	3½	1⅞	1⅞	4-1	4¼	1.36	1.38
1¼	7¼	1½	2½	3¾	2 1/16	2 1/16	4-1⅛	5⅛	1.70	1.72
1½	8	1¾	2⅞	4⅜	2⅜	2⅜	4-1¼	5¾	1.95	1.97
2	9¼	2	3⅝	5	2¾	2¾	8-1⅛	6¾	2.44	2.46
2½	10½	2¼	4⅛	5⅝	3⅛	3⅛	8-1¼	7¾	2.94	2.97
3	12	2⅝	5	6⅝	3⅝	3⅝	8-1⅜	9	3.57	3.60
3½	...	...	...	...	...	...	....	...	...	...
4	14	3	6 3/16	7½	4¼	4¼	8-1⅝	10¾	4.57	4.60
5	16½	3⅝	7 5/16	9	5⅛	5⅛	8-1⅞	12¾	5.66	5.69
6	19	4¼	8½	10¾	6	6	8-2⅛	14½	6.72	6.75
8	21¾	5	10⅝	12½	7	7	12-2⅛	17¼	8.72	8.75
10	26½	6½	12¾	16½	9	9	12-2⅝	21¼	10.88	10.92
12	30	7¼	15	18¼	10	10	12-2⅞	24⅜	12.88	12.92

NOTES:

(1) Always specify bore when ordering

(2) Includes 1/16" raised face in 150# & 300# Standards. Does NOT include ¼" raised face in 400# and heavier standards.

SERIES 150 FLANGE									
PIPE SIZE	FLANGE BOLTS		RAISED FACE				RING JOINT		RING GAP
			LENGTH		GASKET		STUD LENGTH	RING NO.	
	QT'Y	SIZE	STUD	MACH	I.D.	O.D.			
1/2	4	1/2	2 1/4	1 3/4	5/8	1 7/8			
3/4	4	1/2	2 1/4	2	13/16	2 1/4			
1	4	1/2	2 1/2	2	1	2 5/8	3	R-15	5/32
1 1/4	4	1/2	2 1/2	2 1/4	1 3/8	3	3	R-17	–
1 1/2	4	1/2	2 3/4	2 1/4	1 5/8	3 3/8	3 1/4	R-19	–
2	4	5/8	3	2 3/4	2	4 1/8	3 1/2	R-22	–
2 1/2	4	5/8	3 1/4	3	2 1/2	4 7/8	3 3/4	R-25	–
3	4	5/8	3 1/2	3	3	5 3/8	4	R-29	–
3 1/2	8	5/8	3 1/2	3	3 1/2	6 3/8	4	R-33	–
4	8	5/8	3 1/2	3	4	6 7/8	4	R-36	–
5	8	3/4	3 3/4	3 1/4	5	7 3/4	4 1/4	R-40	–
6	8	3/4	3 3/4	3 1/4	6	8 3/4	4 1/4	R-43	–
8	8	3/4	4	3 1/2	8	11	4 1/2	R-48	–
10	12	7/8	4 1/2	3 3/4	10	13 3/8	5	R-52	–
12	12	7/8	4 1/2	4	12	16 1/8	5	R-56	–
14	12	1	5	4 1/4	13 1/4	17 3/4	5 1/2	R-59	1/8
16	16	1	5 1/4	4 1/2	15 1/4	20 1/4	5 3/4	R-64	–
18	16	1 1/8	5 3/4	4 3/4	17 1/4	21 5/8	6 1/4	R-68	–
20	20	1 1/8	6	5 1/4	19 1/4	23 7/8	6 1/2	R-72	–
22	20	1 1/4	6 1/2	5 1/2	21 1/4	26	7	R-80	–
24	20	1 1/4	6 3/4	5 3/4	23 1/4	28 1/4	7 1/4	R-76	–

SERIES 300 FLANGE									
PIPE SIZE	FLANGE BOLTS		RAISED FACE				RING JOINT		RING GAP
			LENGTH		GASKET		STUD LENGTH	RING NO.	
	QT'Y	SIZE	STUD	MACH	I.D.	O.D.			
1/2	4	1/2	2 1/2	2	5/8	2 1/8	3	R-11	1/8
3/4	4	5/8	2 3/4	2 1/2	13/16	2 5/8	3 1/4	R-13	5/32
1	4	5/8	3	2 1/2	1	2 7/8	3 1/2	R-16	–
1 1/4	4	5/8	3	2 3/4	1 3/8	3 1/4	3 1/2	R-18	–
1 1/2	4	3/4	3 1/2	3	1 5/8	3 3/4	4	R-20	–
2	8	5/8	3 1/4	3	2	4 3/8	4	R-23	7/32
2 1/2	8	3/4	3 3/4	3 1/4	2 1/2	5 1/8	4 1/2	R-26	–
3	8	3/4	4	3 1/2	3	5 7/8	4 3/4	R-31	–
3 1/2	8	3/4	4 1/4	3 3/4	3 1/2	6 1/2	5	R-34	–
4	8	3/4	4 1/4	3 3/4	4	7 1/8	5	R-37	–
5	8	3/4	4 1/2	4	5	8 1/2	5 1/4	R-41	–
6	12	3/4	4 3/4	4 1/4	6	9 7/8	5 1/2	R-45	–
8	12	7/8	5 1/4	4 3/4	8	12 1/8	6	R-49	–
10	16	1	6	5 1/4	10	14 1/4	6 3/4	R-53	–
12	16	1 1/8	6 1/2	5 3/4	12	16 5/8	7 1/4	R-57	–
14	20	1 1/8	6 3/4	6	13 1/4	19 1/8	7 1/2	R-61	–
16	20	1 1/4	7 1/4	6 1/2	15 1/4	21 1/4	8	R-65	–
18	24	1 1/4	7 1/2	6 3/4	17	23 1/2	8 1/4	R-69	–
20	24	1 1/4	8	7	19	25 3/4	8 3/4	R-73	–
22	24	1 1/2	8 3/4	7 1/2	21	27 3/4	9 3/4	R-81	–
24	24	1 1/2	9	7 3/4	23	30 1/2	10	R-77	1/4

SERIES 400 FLANGE							
PIPE SIZE	FLANGE BOLTS		STUD BOLT LENGTH			*RING NO.	RING GAP
	QT'Y	SIZE	RAISED FACE	MALE & FEMALE TONGUE & GROOVE	RING JOINT		
4	8	7/8	5 1/4	5	5 1/2	R-37	7/32
5	8	7/8	5 1/2	5 1/4	5 3/4	R-41	–
6	12	7/8	5 3/4	5 1/2	6	R-45	–
8	12	1	6 1/2	6 1/4	6 3/4	R-49	–
10	16	1 1/8	7 1/4	7	7 1/2	R-53	–
12	16	1 1/4	7 3/4	7 1/2	8	R-57	–
14	20	1 1/4	8	7 3/4	8 1/4	R-61	–
16	20	1 3/8	8 1/2	8 1/4	8 3/4	R-65	–
18	24	1 3/8	8 3/4	8 1/2	9	R-69	–
20	24	1 1/2	9 1/2	9 1/4	9 3/4	R-73	–
22	24	1 5/8	10	9 3/4	10 1/2	R-81	3/16
24	24	1 3/4	10 1/2	10 1/4	11	R-77	1/4

*Series 300, 400 & 600 use same ring numbers.

SERIES 600 FLANGE								
PIPE SIZE	FLANGE BOLTS		STUD BOLT LENGTH			*RING NO.	RING GAP
	QT'Y	SIZE	RAISED FACE	MALE & FEMALE TONGUE & GROOVE	RING JOINT		
½	4	½	3	2¾	3	R-11	⅛
¾	4	⅝	3¼	3	3¼	R-13	5/32
1	4	⅝	3½	3¼	3½	R-16	–
1¼	4	⅝	3¾	3½	3¾	R-18	–
1½	4	¾	4	3¾	4	R-20	–
2	8	⅝	4	3¾	4¼	R-23	3/16
2½	8	¾	4½	4¼	4¾	R-26	–
3	8	¾	4¾	4½	5	R-31	–
3½	8	⅞	5¼	5	5½	R-34	–
4	8	⅞	5½	5¼	5¾	R-37	–
5	8	1	6¼	6	6½	R-41	–
6	12	1	6½	6¼	6¾	R-45	–
8	12	1⅛	7½	7¼	7¾	R-49	–
10	16	1¼	8¼	8	8½	R-53	–
12	20	1¼	8½	8¼	8¾	R-57	–
14	20	1⅜	9	8¾	9¼	R-61	–
16	20	1½	9¾	9½	10	R-65	–
18	20	1⅝	10½	10¼	10¾	R-69	–
20	24	1⅝	11¼	11	11½	R-73	–
22	24	1¾	12	11¾	12½	R-81	–
24	24	1⅞	12¾	12½	13¼	R-77	7/32

*Series 300, 400 & 600 use same ring numbers.

SERIES 900 FLANGE							
PIPE SIZE	FLANGE BOLTS		STUD BOLT LENGTH			RING NO.	RING GAP
	QT'Y	SIZE	RAISED FACE	MALE & FEMALE TONGUE & GROOVE	RING JOINT		
3	8	7/8	5½	5¼	5¾	R-31	5/32
4	8	1⅛	6½	6¼	6¾	R-37	–
5	8	1¼	7¼	7	7½	R-41	–
6	12	1⅛	7½	7¼	7½	R-45	–
8	12	1⅜	8½	8¼	8¾	R-49	–
10	16	1⅜	9	8¾	9¼	R-53	–
12	20	1⅜	9¾	9½	10	R-57	–
14	20	1½	10½	10¼	11	R-62	–
16	20	1⅝	11	10¾	11½	R-66	–
18	20	1⅞	12¾	12½	13¼	R-70	3/16
20	20	2	13½	13¼	14	R-74	–
24	20	2½	17	16¾	17¾	R-78	7/32

SERIES 1500 FLANGE							
PIPE SIZE	FLANGE BOLTS		STUD BOLT LENGTH			RING NO.	RING GAP
	QT'Y	SIZE	RAISED FACE	MALE & FEMALE TONGUE & GROOVE	RING JOINT		
1/2	4	3/4	4	3 3/4	4	R-12	5/32
3/4	4	3/4	4 1/4	4	4 1/4	R-14	–
1	4	7/8	4 3/4	4 1/2	4 3/4	R-16	–
1 1/4	4	7/8	4 3/4	4 1/2	4 3/4	R-18	–
1 1/2	4	1	5 1/4	5	5 1/4	R-20	–
2	8	7/8	5 1/2	5 1/4	5 3/4	R-24	1/8
2 1/2	8	1	6	5 3/4	6 1/4	R-27	–
3	8	1 1/8	6 3/4	6 1/2	7	R-35	–
4	8	1 1/4	7 1/2	7 1/4	7 3/4	R-39	–
5	8	1 1/2	9 1/2	9 1/4	9 3/4	R-44	–
6	12	1 3/8	10	9 3/4	10 1/4	R-46	–
8	12	1 5/8	11 1/4	11	11 3/4	R-50	5/32
10	12	1 7/8	13 1/4	13	13 1/2	R-54	–
12	16	2	14 3/4	14 1/2	15 1/4	R-58	3/16
14	16	2 1/4	16	15 3/4	16 3/4	R-63	7/32
16	16	2 1/2	17 1/2	17 1/4	18 1/2	R-67	5/16
18	16	2 3/4	19 1/4	19	20 1/4	R-71	–
20	16	3	21	20 3/4	22 1/4	R-75	3/8
24	16	3 1/2	24	23 3/4	25 1/2	R-79	7/16

SERIES 2500 FLANGE

PIPE SIZE	FLANGE BOLTS		STUD BOLT LENGTH			RING NO.	RING GAP
	QT'Y	SIZE	RAISED FACE	MALE & FEMALE TONGUE & GROOVE	RING JOINT		
½	4	¾	4¾	4½	4¾	R-13	5/32
¾	4	¾	4¾	4½	4¾	R-16	–
1	4	⅞	5¼	5	5¼	R-18	–
1¼	4	1	5¾	5½	6	R-21	⅛
1½	4	1⅛	6½	6¼	6¾	R-23	–
2	8	1	6¾	6½	7	R-26	–
2½	8	1⅛	7½	7¼	7¾	R-28	–
3	8	1¼	3½	8¼	8¾	R-32	–
4	8	1½	9¾	9½	10¼	R-38	5/32
5	8	1¾	11½	11¼	12¼	R-42	–
6	8	2	13½	13¼	14	R-47	–
8	12	2	15	14¾	15½	R-51	3/16
10	12	2½	19	18¾	20	R-55	¼
12	12	2¾	21	20¾	22	R-60	5/16

WRENCH SIZES

BOLT DIAM	WRENCH SIZE	BOLT DIAM	WRENCH SIZE
1/2	7/8	1 5/8	2 9/16
5/8	1 1/16	1 3/4	2 3/4
3/4	1 1/4	1 7/8	2 15/16
7/8	1 7/16	2	3 1/8
1	1 5/8	2 1/4	3 1/2
1 1/8	1 13/16	2 1/2	3 7/8
1 1/4	2	2 3/4	4 1/4
1 3/8	2 3/16	3	4 5/8
1 1/2	2 3/8	3 1/2	5 3/8

DRILL SIZES FOR PIPE TAPS

Size of Tap in Inches	No. of Threads Per Inch	Diam. of Drill	Size of Tap in Inches	No. of Threads Per Inch	Diam. of Drill
1/8	27	11/32	2	11 1/2	2 3/16
1/4	18	7/16	2 1/2	8	2 9/16
3/8	18	37/64	3	8	3 3/16
1/2	14	23/32	3 1/2	8	3 11/16
3/4	14	59/64	4	8	4 3/16
1	11 1/2	1 5/32	4 1/2	8	4 3/4
1 1/4	11 1/2	1 1/2	5	8	5 5/16
1 1/2	11 1/2	1 49/64	6	8	6 5/16

TAP AND DRILL SIZES
(American Standard Coarse)

Size of Drill	Size of Tap	Threads Per Inch	Size of Drill	Size of Tap	Threads Per Inch
7	1/4	20	49/64	7/8	9
F	5/16	18	53/64	15/16	9
5/16	3/8	16	7/8	1	8
U	7/16	14	63/64	1 1/8	7
27/64	1/2	13	1 7/64	1 1/4	7
31/64	9/16	12	1 13/64	1 3/8	6
17/32	5/8	11	1 11/32	1 1/2	6
19/32	11/16	11	1 29/64	1 5/8	5 1/2
21/32	3/4	10	1 11/16	1 7/8	5
23/32	13/16	10	1 11/16	1 7/8	4 1/2
			1 25/32	2	4 1/2

BOLT CHART FOR 300 LB. & 400 LB. ORIFICE FLANGES

NOM PIPE SIZE	FLANGE BOLTS		300 LB. ORIFICE STUD LENGTH		NOM PIPE SIZE	FLANGE BOLTS		400 LB. ORIFICE STUD LENGTH	
	QT'Y	SIZE	RF	RTJ		QT'Y	SIZE	RF	RTJ
1	4	5/8	4	4 3/4	NOTE: ①				
1 1/4	4	5/8	4	4 3/4					
1 1/2	4	3/4	4 1/4	5	4	8	7/8	5 1/2	6
2	8	5/8	4	4 3/4	5	8	7/8	5 3/4	6 1/4
2 1/2	8	3/4	4 1/4	5	6	12	7/8	6 1/4	6 1/2
3	8	3/4	4 1/4	5	8	12	1	6 3/4	7 1/4
4	8	3/4	4 1/4	5	10	16	1 1/8	7 5/8	8
5	8	3/4	4 1/4	5 1/2	12	16	1 1/4	8	8 1/2
6	12	3/4	4 1/4	5 1/2	14	20	1 1/4	8 1/4	9
8	12	7/8	4 1/2	6	16	20	1 3/8	8 3/4	9 1/4
10	16	1	5 1/2	6 1/2	18	24	1 3/8	9 1/4	9 1/2
12	16	1 1/8	5 1/2	7	20	24	1 1/2	9 3/4	10 1/4
14	20	1 1/8	6	7	22	24	1 5/8	10 1/2	10 3/4
16	20	1 1/4	6 1/2	8	24	24	1 3/4	11	11 1/4
18	24	1 1/4	6 1/2	8					
20	24	1 1/4	7	8					
22	24	1 1/2	7 1/2	9					
24	24	1 1/2	7 1/2	9					

NOTE:

① For sizes 1" through 3" use 300 LB. orifice flanges.

BOLT CHART FOR 600 LB., 900 LB. & 1500 LB. ORIFICE FLANGES

NOM SIZE PIPE	FLANGE BOLTS		600 LB. ORIFICE STUD LENGTH	
	QT'Y	SIZE	RF	RTJ
NOTE: ①				
4	8	7/8	5 3/4	6 1/4
5	8	1	6 1/2	7
6	12	1	6 3/4	7 1/4
8	12	1 1/8	7 5/8	8 1/4
10	16	1 1/4	8 1/2	9
12	20	1 1/4	9	9 1/4
14	20	1 3/8	9 1/4	9 3/4
16	20	1 1/2	10	10 1/2
18	20	1 5/8	11	11 1/4
20	24	1 5/8	11 3/4	12
22	24	1 3/4	12 1/2	13
24	24	1 7/8	13 1/4	13 1/2
NOTE: ②			900 LB. ORIFICE	
3	8	7/8	5 3/4	6 1/4
4	8	1 1/8	7	7 1/4
5	8	1 1/4	7 1/2	8
6	12	1 1/8	7 5/8	8 1/4
8	12	1 3/8	9	9 1/4
10	16	1 3/8	9 1/4	9 3/4
12	20	1 3/8	10	10 3/4

NOM SIZE PIPE	FLANGE BOLTS		1500 LB. ORIFICE STUD LENGTH	
	QT'Y	SIZE	RF	RTJ
NOTE: ③				
1	4	7/8	5 1/2	5 3/4
1 1/4	4	7/8	5 1/2	5 3/4
1 1/2	4	1	5 3/4	6
2	8	7/8	5 3/4	6 1/4
2 1/2	8	1	6 1/4	6 3/4
3	8	1 1/8	7	7 5/8
4	8	1 1/4	7 3/4	8 1/4
5	8	1 1/2	9 3/4	10 1/4
6	12	1 3/8	10 1/4	11
8	12	1 5/8	11 1/2	12 1/2
10	12	1 7/8	13 1/4	14 1/4
12	16	2	14 3/4	16

NOTES:

① For sizes 1'' through 3'' use 300 LB. orifice flanges.

② For sizes 1'' through 2½'' use 1500 LB. orifice flanges.

③ 2500 LB. orifice flanges are also available.

OUTSIDE AND INSIDE DIAMETERS OF PIPE AND BORES FOR WELDING NECK AND SOCKET WELD FLANGES.

Nom. Pipe Size	Outside Diam.	Light Wall●	Sched. 20	Sched. 30	Std. Wall.	Sched. 40	Sched. 60	Extra Strong	Sched. 80	Sched. 100	Sched. 120	Sched. 140	Sched. 160	Double Extra Strong
½	0.840	.674	...	...	0.622	0.622	...	0.546	0.546	...	...	...	0.464	0.252
¾	1.050	.884	...	...	0.824	0.824	...	0.742	0.742	...	...	...	0.612	0.434
1	1.315	1.097	...	...	1.049	1.049	...	0.957	0.957	...	...	...	0.815	0.599
1¼	1.660	1.442	...	...	1.380	1.380	...	1.278	1.278	...	...	...	1.160	0.896
1½	1.900	1.682	...	...	1.610	1.610	...	1.500	1.500	...	...	...	1.338	1.100
2	2.375	2.157	...	...	2.067	2.067	...	1.939	1.939	...	...	...	1.687	1.503
2½	2.875	2.635	...	...	2.469	2.469	...	2.323	2.323	...	...	...	2.125	1.771
3	3.500	3.260	...	...	3.068	3.068	...	2.900	2.900	...	...	...	2.624	2.300
3½	4.000	3.760	...	...	3.548	3.548	...	3.364	3.364	...	...	...	...	2.728
4	4.500	4.260	...	...	4.026	4.026	...	3.826	3.826	...	3.624	...	3.438	3.152
5	5.563	5.295	...	...	5.047	5.047	...	4.813	4.813	...	4.563	...	4.313	4.063
6	6.625	6.357	...	...	6.065	6.065	...	5.761	5.761	...	5.501	...	5.187	4.897
8	8.625	8.329	8.125	8.071	7.981	7.981	7.813	7.625	7.625	7.437	7.187	7.001	6.813	6.875
10	10.750	10.420	10.250	10.136	10.020	10.020	9.750	9.750	9.562	9.312	9.062	8.750	8.500	8.750
12	12.750	12.390	12.250	12.090	12.000	11.938	11.626	11.750	11.374	11.062	10.750	10.500	10.126	10.750
14	14.000	13.500	13.376	13.250	13.250	13.124	12.812	13.000	12.500	12.124	11.814	11.500	11.188	...
16	16.000	15.500	15.376	15.250	15.250	15.000	14.688	15.000	14.312	13.938	13.564	13.124	12.812	...
18	18.000	17.500	17.376	17.124	17.250	16.876	16.500	17.000	16.124	15.688	15.250	14.876	14.438	...
20	20.000	19.500	19.250	19.000	19.250	18.812	18.376	19.000	17.938	17.438	17.000	16.500	16.062	...
24	24.000	23.500	23.250	22.876	23.250	22.624	22.062	23.000	21.562	20.938	20.376	19.876	19.312	...
30	30.000	29.376	29.000	28.750	29.250	...	...	29.000	...	...	...	...	...	...
36	36.000	35.376	35.000	34.750	35.250	34.500	...	35.000	...	...	...	...	...	...
42	42.000	...	...	...	41.250	...	...	41.000	...	...	...	...	...	...

NOTE: ● Light wall diameters are the same as stainless steel Schedule 10s — in sizes thru 12" and to Schedule 10 in sizes 14" and larger.

COMPARISON CHART FOR PACKING AND GASKET MATERIALS

COMPANY	GASKET MATERIAL				PACKING			
ANCHOR	424	425	450	4250	105	103	888	
BELMONT	590	590	584	6735	30-C	189-C	6504	6106
CHESTERTON	210	235	270	260	350		318	340
COLLINS	920		3004		1251	1641		47-G
DURABLA	√							
JOHN CRANE	334		891	2112	800	804-D	896	1810
GARLOCK	7735	900	7228	7705	150	176	230	237
GREENE, TWEED	2900	2905	2910	2970	2206	1130		
HERCULES	565		562	570	101	127	191	138
JOHNS MANVILLE	60	61	76	84	166	731	2018	18
RAYBESTOS MANHATTAN	670	501	K-68	1307	121-C	376-C	380	
SOUTHEASTERN PRODUCTS	300		240	400	151-RB		845	168
STERLING	415	417						

NOTE: Some of these gasket materials may be ordered ungraphited or with one or both sides graphited. Refer to suppliers catalog for more information.

USA STANDARD

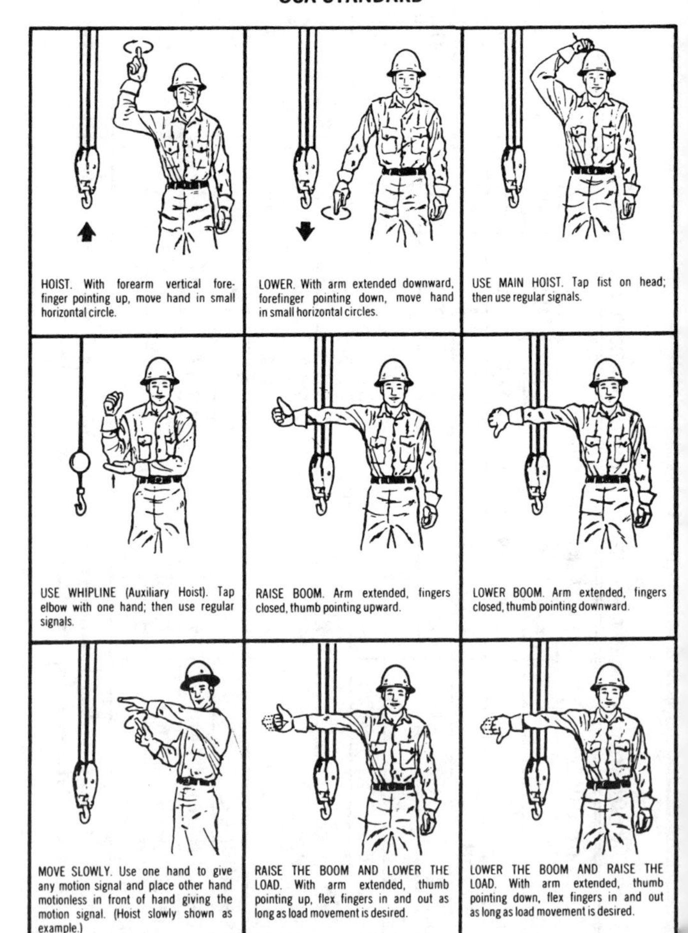

FIG. 1 STANDARD HAND SIGNALS FOR CONTROLLING CRANE OPERATIONS

From B30.5-1968 Crawler, Locomotive, and Truck Cranes. With permission of the American Society of Mechanical Engineers, the publisher. New York, N.Y. 10017.

FIG. 2 STANDARD HAND SIGNALS FOR CONTROLLING CRANE OPERATIONS

From B30.5-1968 Crawler, Locomotive, and Truck Cranes With permission of The American Society of Mechanical Engineers, The publisher. New York, N.Y. 10017.

HOIST. With forearm vertical, forefinger pointing up, move hand in small horizontal circle.

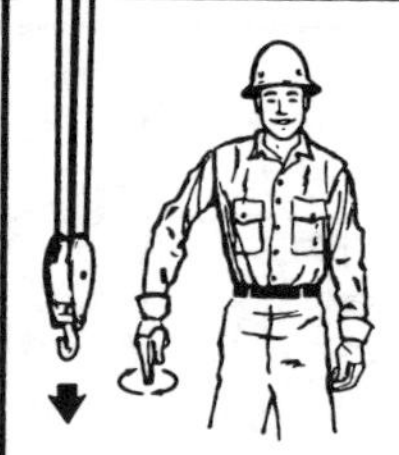

LOWER. With arm extended downward, forefinger pointing down, move hand in small horizontal circle.

BRIDGE TRAVEL. Arm extended forward, hand open and slightly raised, make pushing motion in direction of travel.

TROLLEY TRAVEL. Palm up, fingers closed, thumb pointing in direction of motion, jerk hand horizontally.

STOP. Arm extended, palm down, hold position rigidly.

EMERGENCY STOP. Arm extended, palm down, move hand rapidly right and left.

MULTIPLE TROLLEYS. Hold up one finger for block marked "1" and two fingers for block marked "2." Regular signals follow.

MOVE SLOWLY. Use one hand to give any motion signal and place other hand motionless in front of hand giving the motion signal. (Hoist slowly shown as example.)

POWER IS DISCONNECTED. Crane operator spreads both hands apart — palms up.

FIG. 3 STANDARD HAND SIGNALS FOR CONTROLLING OVERHEAD AND GANTRY CRANES

From B30.2-1967 Overhead and Gantry Cranes. With permission of The American Society of Mechanical Engineers, the publisher. New York, N.Y. 10017.

USEFUL INFORMATION

TOTAL EXPANSION PER 100 FEET

1. Steel pipe equals (final temperature minus starting temperature) X .00804
2. Copper and brass pipe equals (final temperature minus starting temperature) X .0114
3. For lengths less than 100 feet move decimal point two places to the left. Examples: 72 feet equal .72, 172 feet equal 1.72 X above results for 100 feet

PIPE BENDS

1. The **minimum** radius is most often given as 6 X pipe size.
2. The amount of pipe required for a bend equals; The radius X the degrees X .01745.

THE CIRCLE

1. Circumference equals diameter X 3.1416.
2. Diameter equals circumference X .31831.
3. Arc length equals radius X degrees X .01745.
4. Degrees of arc equals length divided by (radius X .01745)
5. Radius of arc equals length divided by (degrees X .01745).

HUNDREDS OF FEET

1. To change hundredths of feet to inches multiply by 12.
2. To change inches to hundredths of feet multiply by .0833.

INCHES AND FEET CONVERTED TO MILLIMETERS

1"	25.4	7"	177.8	1'	304.8	7'	2133.6
2"	50.8	8"	203.2	2'	609.6	8'	2438.4
3"	76.2	9"	228.6	3'	914.4	9'	2743.2
4"	101.6	10"	254.0	4'	1219.2	10'	3048.0
5"	127.0	11"	279.4	5'	1524.0	11'	3352.8
6"	152.4	12"	304.8	6'	1828.8	12'	3657.6

MOVE ABOVE DECIMAL POINT TO LEFT
FOR: CENTIMETERS ONE SPACE
FOR: DECIMETERS TWO SPACES
FOR: METERS THREE SPACES

SEE PAGE 126 FOR FRACTIONS TO MILLIMETERS

CONVERSION CONSTANTS

TO CHANGE	TO	MULTIPLY BY
Inches	Feet	0.0833
Inches	Millimeters	25.4
Feet	Inches	12
Feet	Yards	0.3333
Yards	Feet	3
Square inches	Square feet	0.00694
Square feet	Square inches	144
Square feet	Square yards	0.11111
Square yards	Square feet	9
Cubic inches	Cubic feet	0.00058
Cubic feet	Cubic inches	1728
Cubic feet	Cubic yards	0.03703
Cubic yards	Cubic feet	27
Cubic inches	Gallons	0.00433
Cubic feet	Gallons	7.48
Gallons	Cubic inches	231
Gallons	Cubic feet	0.1337
Gallons	Pounds of water	8.33
Pounds of water	Gallons	0.12004
Ounces	Pounds	0.0625
Pounds	Ounces	16
Inches of water	Pounds per square inch	0.0361
Inches of water	Inches of mercury	0.0735
Inches of water	Ounces per square inch	0.578
Inches of water	Pounds per square foot	5.2
Inches of mercury	Inches of water	13.6
Inches of mercury	Feet of water	1.1333
Inches of mercury	Pounds per square inch	0.4914
Ounces per square inch	Inches of mercury	0.127
Ounces per square inch	Inches of water	1.733
Pounds per square inch	Inches of water	27.72
Pounds per square inch	Feet of water	2.310
Pounds per square inch	Inches of mercury	2.04
Pounds per square inch	Atmospheres	0.0681
Feet of water	Pounds per square inch	0.434
Feet of water	Pounds per square foot	62.5
Feet of water	Inches of mercury	0.8824
Atmospheres	Pounds per square inch	14.696
Atmospheres	Inches of mercury	29.92
Atmospheres	Feet of water	34
Long tons	Pounds	2240
Short tons	Pounds	2000
Short tons	Long tons	0.89285

DECIMAL EQUIVALENTS

Fraction			Decimal	Millimeters	Fraction			Decimal	Millimeters
		1/64	.01563	0.397			33/64	.51563	13.097
	1/32		.03125	0.794		17/32		.53125	13.494
		3/64	.04688	1.191			35/64	.54688	13.891
1/16			.0625	1.588	9/16			.5625	14.288
		5/64	.07813	1.984			37/64	.57813	14.684
	3/32		.09375	2.381		19/32		.59375	15.081
		7/64	.10938	2.778			39/64	.60938	15.478
1/8			.125	3.175	5/8			.625	15.875
		9/64	.14063	3.572			41/64	.64063	16.272
	5/32		.15625	3.969		21/32		.65625	16.669
		11/64	.17188	4.366			43/64	.67188	17.066
3/16			.1875	4.763	11/16			.6875	17.463
		13/64	.20313	5.159			45/64	.70313	17.859
	7/32		.21875	5.556		23/32		.71875	18.256
		15/64	.23438	5.953			47/64	.73438	18.653
1/4			.250	6.350	3/4			.750	19.050
		17/64	.26563	6.747			49/64	.76563	19.447
	9/32		.28125	7.144		25/32		.78125	19.844
		19/64	.29688	7.541			51/64	.79688	20.241
5/16			.3125	7.938	13/16			.8125	20.638
		21/64	.32813	8.334			53/64	.82813	21.034
	11/32		.34375	8.731		27/32		.84375	21.431
		23/64	.35938	9.128			55/64	.85938	21.828
3/8			.375	9.525	7/8			.875	22.225
		25/64	.39063	9.922			57/64	.89063	22.622
	13/32		.40625	10.319		29/32		.90625	23.019
		27/64	.42188	10.716			59/64	.92188	23.416
7/16			.4375	11.113	15/16			.9375	23.813
		29/64	.45313	11.509			61/64	.95313	24.209
	15/32		.46875	11.906		31/32		.96875	24.606
		31/64	.48438	12.303			63/64	.98438	25.003
1/2			.500	12.700	1			1.00000	25.400

WIRE ROPE

Safe Load in Pounds for New Wire Rope
6 Strands of 7 Wires, Hemp Center

DIAM. IN INCHES	SAFE LOAD POUNDS	DIAM IN INCHES	SAFE LOAD POUNDS	DIAM. IN INCHES	SAFE LOAD POUNDS
1/4	940	9/16	4,500	1-1/8	17,400
5/16	1,400	5/8	5,500	1-1/4	21,200
3/8	2,000	3/4	7,900	1-3/8	25,400
7/16	2,700	7/8	10,700	1-1/2	30,000
1/2	3,600	1	13,900		

WHEN ROPES ARE GALVANIZED
DEDUCT 10% FROM STRENGTH SHOWN ABOVE

MANILA ROPE

Safe Load for New Manila Rope — 3 Strand
Safety Factor - 7

DIAM. IN INCHES	SAFE LOAD POUNDS	DIAM. IN INCHES	SAFE LOAD POUNDS	DIAM. IN INCHES	SAFE LOAD POUNDS
1/4	85	3/4	780	1-1/4	1,900
3/8	185	13/16	920	1-1/2	2,640
1/2	360	1	1,280	1-13/16	3,700
				2	4,400

RULE OF THUMB

Open Eye Hook Safe load in tons is diameter of eye in inches squared.

2" hook, 2x2 = 4 Tons.

Shackle Safe load in tons is diameter of a pin in one-fourth inches (1/4") squared and divided by three (3).

1/2" diameter = 2 quarters

$$\frac{2 \times 2}{3} = 1\text{-}1/3 \text{ tons or } 2{,}667 \text{ pounds.}$$

Chains Safe load in tons is six (6) times the diameter of chain stock in inches squared.

1/2" diameter chain stock

1/2 x 1/2 x 6 = 1-1/2 tons or 3,000 lbs.

DECIMALS OF A FOOT

INCH	0″	1″	2″	3″	4″	5″	6″	7″	8″	9″	10″	11″
0	0	.0833	.1667	.2500	.3333	.4167	.5000	.5833	.6667	.7500	.8333	.9167
1/16	.0052	.0885	.1719	.2552	.3385	.4219	.5052	.5885	.6719	.7552	.8385	.9219
1/8	.0104	.0938	.1771	.2604	.3438	.4271	.5104	.5938	.6771	.7604	.8438	.9271
3/16	.0156	.0990	.1823	.2656	.3490	.4323	.5156	.5990	.6823	.7656	.8490	.9323
1/4	.0208	.1042	.1875	.2708	.3542	.4375	.5208	.6042	.6875	.7708	.8542	.9375
5/16	.0260	.1094	.1927	.2760	.3594	.4427	.5260	.6094	.6927	.7760	.8594	.9427
3/8	.0313	.1146	.1979	.2812	.3646	.4479	.5313	.6146	.6979	.7813	.8646	.9479
7/16	.0365	.1198	.2031	.2865	.3698	.4531	.5365	.6198	.7031	.7865	.8698	.9531
1/2	.0417	.1250	.2083	.2917	.3750	.4583	.5417	.6250	.7083	.7917	.8750	.9583
9/16	.0469	.1302	.2135	.2969	.3802	.4635	.5469	.6302	.7135	.7969	.8802	.9635
5/8	.0521	.1354	.2188	.3021	.3854	.4688	.5521	.6354	.7188	.8021	.8854	.9688
11/16	.0573	.1406	.2240	.3073	.3906	.4740	.5573	.6406	.7240	.8073	.8906	.9740
3/4	.0625	.1458	.2292	.3125	.3958	.4792	.5625	.6458	.7292	.8125	.8958	.9792
13/16	.0677	.1510	.2344	.3177	.4010	.4844	.5677	.6510	.7344	.8177	.9010	.9844
7/8	.0729	.1563	.2396	.3229	.4063	.4896	.5729	.6563	.7396	.8229	.9063	.9896
15/16	.0781	.1615	.2448	.3281	.4115	.4948	.5781	.6615	.7448	.8281	.9115	.9948
1	.0833	.1667	.2500	.3333	.4167	.5000	.5833	.6667	.7500	.8333	.9167	1.0000

COTANGENT
SINE
TANGENT
COSINE
RADIUS=1
SECANT
COSECANT

MINUTES CONVERTED TO DECIMALS OF A DEGREE

MIN.	DEG.	MIN.	DEG.	MIN.	DEG.	MIN.	DEG.	MIN.	DEG.	MIN.	DEG.
1	.0166	11	.1833	21	.3500	31	.5166	41	.6833	51	.8500
2	.0333	12	.2000	22	.3666	32	.5333	42	.7000	52	.8666
3	.0500	13	.2166	23	.3833	33	.5500	43	.7166	53	.8833
4	.0666	14	.2333	24	.4000	34	.5666	44	.7333	54	.9000
5	.0833	15	.2500	25	.4166	35	.5833	45	.7500	55	.9166
6	.1000	16	.2666	26	.4333	36	.6000	46	.7666	56	.9333
7	.1166	17	.2833	27	.4500	37	.6166	47	.7833	57	.9500
8	.1333	18	.3000	28	.4666	38	.6333	48	.8000	58	.9666
9	.1500	19	.3166	29	.4833	39	.6500	49	.8166	59	.9833
10	.1666	20	.3333	30	.5000	40	.6666	50	.8333	60	1.0000

FORMULAS FOR FINDING FUNCTIONS OF ANGLES

$$\frac{\text{Side opposite}}{\text{Hypotenuse}} = \text{SINE}$$

$$\frac{\text{Side adjacent}}{\text{Hypotenuse}} = \text{COSINE}$$

$$\frac{\text{Side opposite}}{\text{Side adjacent}} = \text{TANGENT}$$

$$\frac{\text{Side adjacent}}{\text{Side opposite}} = \text{COTANGENT}$$

$$\frac{\text{Hypotenuse}}{\text{Side adjacent}} = \text{SECANT}$$

$$\frac{\text{Hypotenuse}}{\text{Side opposite}} = \text{COSECANT}$$

FORMULAS FOR FINDING THE LENGTH OF SIDES FOR RIGHT-ANGLE TRIANGLES WHEN AN ANGLE AND SIDE ARE KNOWN

Length of side opposite	Hypotenuse × Sine Hypotenuse ÷ Cosecant Side adjacent × Tangent Side adjacent ÷ Cotangent
Length of side adjacent	Hypotenuse × Cosine Hypotenuse ÷ Secant Side opposite × Cotangent Side opposite ÷ Tangent
Length of Hypotenuse	Side opposite × Cosecant Side opposite ÷ Sine Side adjacent × Secant Side adjacent ÷ Cosine

0°

M	Sine	Cosine	Tan.	Cotan.	Secant	Cosec.	M
0	.00000	1.0000	.00000	Infinite	1.0000	Infinite	60
1	.00029	.0000	.00029	3437.7	.0000	3437.7	59
2	.00058	.0000	.00058	1718.9	.0000	1718.9	58
3	.00087	.0000	.00087	1145.9	.0000	1145.9	57
4	.00116	.0000	.00116	859.44	.0000	859.44	56
5	.00145	1.0000	.00145	687.55	1.0000	687.55	55
6	.00174	.0000	.00174	572.96	.0000	572.96	54
7	.00204	.0000	.00204	491.11	.0000	491.11	53
8	.00233	.0000	.00233	429.72	.0000	429.72	52
9	.00262	.0000	.00262	381.97	.0000	381.97	51
10	.00291	.99999	.00291	343.77	1.0000	343.77	50
11	.00320	.99999	.00320	312.52	.0000	312.52	49
12	.00349	.99999	.00349	286.48	.0000	286.48	48
13	.00378	.99999	.00378	264.44	.0000	264.44	47
14	.00407	.99999	.00407	245.55	.0000	245.55	46
15	.00436	.99999	.00436	229.18	1.0000	229.18	45
16	.00465	.99999	.00465	214.86	.0000	214.86	44
17	.00494	.99999	.00494	202.22	.0000	202.22	43
18	.00524	.99999	.00524	190.98	.0000	190.99	42
19	.00553	.99998	.00553	180.93	.0000	180.93	41
20	.00582	.99998	.00582	171.88	1.0000	171.89	40
21	.00611	.99998	.00611	163.70	.0000	163.70	39
22	.00640	.99998	.00640	156.26	.0000	156.26	38
23	.00669	.99998	.00669	149.46	.0000	149.47	37
24	.00698	.99997	.00698	143.24	.0000	143.24	36
25	.00727	.99997	.00727	137.51	1.0000	137.51	35
26	.00756	.99997	.00756	132.22	.0000	132.22	34
27	.00785	.99997	.00785	127.32	.0000	127.32	33
28	.00814	.99997	.00814	122.77	.0000	122.78	32
29	.00843	.99996	.00844	118.54	.0000	118.54	31
30	.00873	.99996	.00873	114.59	1.0000	114.59	30
31	.00902	.99996	.00902	110.89	.0000	110.90	29
32	.00931	.99996	.00931	107.43	.0000	107.43	28
33	.00960	.99995	.00960	104.17	.0000	104.17	27
34	.00989	.99995	.00989	101.11	.0000	101.11	26
35	.01018	.99995	.01018	98.218	1.0000	98.223	25
36	.01047	.99994	.01047	95.489	.0000	95.495	24
37	.01076	.99994	.01076	92.908	.0000	92.914	23
38	.01105	.99994	.01105	90.463	.0001	90.469	22
39	.01134	.99993	.01134	88.143	.0001	88.149	21
40	.01163	.99993	.01164	85.940	1.0001	85.946	20
41	.01193	.99993	.01193	83.843	.0001	83.849	19
42	.01222	.99992	.01222	81.847	.0001	81.853	18
43	.01251	.99992	.01251	79.943	.0001	79.950	17
44	.01280	.99992	.01280	78.126	.0001	78.133	16
45	.01309	.99991	.01309	76.390	1.0001	76.396	15
46	.01338	.99991	.01338	74.729	.0001	74.736	14
47	.01367	.99991	.01367	73.139	.0001	73.146	13
48	.01396	.99990	.01396	71.615	.0001	71.622	12
49	.01425	.99990	.01425	70.153	.0001	70.160	11
50	.01454	.99989	.01454	68.750	1.0001	68.757	10
51	.01483	.99989	.01484	67.402	.0001	67.409	9
52	.01512	.99988	.01513	66.105	.0001	66.113	8
53	.01542	.99988	.01542	64.858	.0001	64.866	7
54	.01571	.99988	.01571	63.657	.0001	63.664	6
55	.01600	.99987	.01600	62.499	1.0001	62.507	5
56	.01629	.99987	.01629	61.383	.0001	61.391	4
57	.01658	.99987	.01658	60.306	.0001	60.314	3
58	.01687	.99986	.01687	59.266	.0001	59.274	2
59	.01716	.99985	.01716	58.261	.0001	58.270	1
60	.01745	.99985	.01745	57.290	1.0001	57.299	0
M	Cosine	Sine	Cotan.	Tan.	Cosec.	Secant	M

89°

1°

M	Sine	Cosine	Tan.	Cotan.	Secant	Cosec.	M
0	.01745	.99985	.01745	57.290	1.0001	57.299	60
1	.01774	.99984	.01775	56.350	.0001	56.359	59
2	.01803	.99984	.01804	55.441	.0001	55.450	58
3	.01832	.99983	.01833	54.561	.0002	54.570	57
4	.01861	.99983	.01862	53.708	.0002	53.718	56
5	.01891	.99982	.01891	52.882	1.0002	52.891	55
6	.01920	.99981	.01920	52.081	.0002	52.090	54
7	.01949	.99981	.01949	51.303	.0002	51.313	53
8	.01978	.99980	.01978	50.548	.0002	50.558	52
9	.02007	.99980	.02007	49.816	.0002	49.826	51
10	.02036	.99979	.02036	49.104	1.0002	49.114	50
11	.02065	.99979	.02066	48.412	.0002	48.422	49
12	.02094	.99978	.02095	47.739	.0002	47.750	48
13	.02123	.99977	.02124	47.085	.0002	47.096	47
14	.02152	.99977	.02153	46.449	.0002	46.460	46
15	.02181	.99976	.02182	45.829	1.0002	45.840	45
16	.02210	.99975	.02211	45.226	.0002	45.237	44
17	.02240	.99975	.02240	44.638	.0002	44.650	43
18	.02269	.99974	.02269	44.066	.0002	44.077	42
19	.02298	.99974	.02298	43.508	.0003	43.520	41
20	.02326	.99973	.02327	42.964	1.0003	42.976	40
21	.02356	.99972	.02357	42.433	.0003	42.445	39
22	.02385	.99971	.02386	41.916	.0003	41.928	38
23	.02414	.99971	.02415	41.410	.0003	41.423	37
24	.02443	.99970	.02444	40.917	.0003	40.930	36
25	.02472	.99969	.02473	40.436	1.0003	40.448	35
26	.02501	.99969	.02502	39.965	.0003	39.978	34
27	.02530	.99968	.02531	39.506	.0003	39.518	33
28	.02559	.99967	.02560	39.057	.0003	39.069	32
29	.02589	.99966	.02589	38.618	.0003	38.631	31
30	.02618	.99966	.02618	38.188	1.0003	38.201	30
31	.02647	.99965	.02648	37.769	.0003	37.782	29
32	.02676	.99964	.02677	37.358	.0003	37.371	28
33	.02705	.99963	.02706	36.956	.0004	36.969	27
34	.02734	.99963	.02735	36.563	.0004	36.576	26
35	.02763	.99962	.02764	36.177	1.0004	36.191	25
36	.02792	.99961	.02793	35.800	.0004	35.814	24
37	.02821	.99960	.02822	35.431	.0004	35.445	23
38	.02850	.99959	.02851	35.069	.0004	35.084	22
39	.02879	.99958	.02880	34.715	.0004	34.729	21
40	.02908	.99958	.02910	34.368	1.0004	34.382	20
41	.02937	.99957	.02939	34.027	.0004	34.042	19
42	.02967	.99956	.02968	33.693	.0004	33.708	18
43	.02996	.99955	.02997	33.366	.0004	33.381	17
44	.03025	.99954	.03026	33.045	.0004	33.060	16
45	.03054	.99953	.03055	32.730	1.0005	32.745	15
46	.03083	.99952	.03084	32.421	.0005	32.437	14
47	.03112	.99951	.03113	32.118	.0005	32.134	13
48	.03141	.99951	.03143	31.820	.0005	31.836	12
49	.03170	.99950	.03172	31.528	.0005	31.544	11
50	.03199	.99949	.03201	31.241	1.0005	31.257	10
51	.03228	.99948	.03230	30.960	.0005	30.976	9
52	.03257	.99947	.03259	30.683	.0005	30.699	8
53	.03286	.99946	.03288	30.411	.0005	30.428	7
54	.03315	.99945	.03317	30.145	.0005	30.161	6
55	.03344	.99944	.03346	29.882	1.0005	29.899	5
56	.03374	.99943	.03375	29.624	.0006	29.641	4
57	.03403	.99942	.03405	29.371	.0006	29.388	3
58	.03432	.99941	.03434	29.122	.0006	29.139	2
59	.03461	.99940	.03463	28.877	.0006	28.894	1
60	.03490	.99939	.03492	28.636	1.0006	28.654	0
M	Cosine	Sine	Cotan.	Tan.	Cosec.	Secant	M

88°

2°

M	Sine	Cosine	Tan.	Cotan.	Secant	Cosec.	M
0	.03490	.99939	.03492	28.636	1.0006	28.654	60
1	.03519	.99938	.03521	28.399	.0006	28.417	59
2	.03548	.99937	.03550	28.166	.0006	28.184	58
3	.03577	.99936	.03579	27.937	.0006	27.955	57
4	.03606	.99935	.03608	27.712	.0006	27.730	56
5	.03635	.99934	.03638	27.490	1.0007	27.508	55
6	.03664	.99933	.03667	27.271	.0007	27.290	54
7	.03693	.99932	.03696	27.056	.0007	27.075	53
8	.03722	.99931	.03725	26.845	.0007	26.864	52
9	.03751	.99930	.03754	26.637	.0007	26.655	51
10	.03781	.99928	.03783	26.432	1.0007	26.450	50
11	.03810	.99927	.03812	26.230	.0007	26.249	49
12	.03839	.99926	.03842	26.031	.0007	26.050	48
13	.03868	.99925	.03871	25.835	.0007	25.854	47
14	.03897	.99924	.03900	25.642	.0008	25.661	46
15	.03926	.99923	.03929	25.452	1.0008	25.471	45
16	.03955	.99922	.03958	25.264	.0008	25.284	44
17	.03984	.99921	.03987	25.080	.0008	25.100	43
18	.04013	.99919	.04016	24.898	.0008	24.918	42
19	.04042	.99918	.04045	24.718	.0008	24.739	41
20	.04071	.99917	.04075	24.542	1.0008	24.562	40
21	.04100	.99916	.04104	24.367	.0008	24.388	39
22	.04129	.99915	.04133	24.196	.0008	24.216	38
23	.04158	.99913	.04162	24.026	.0009	24.047	37
24	.04187	.99912	.04191	23.859	.0009	23.880	36
25	.04217	.99911	.04220	23.694	1.0009	23.716	35
26	.04246	.99910	.04249	23.532	.0009	23.553	34
27	.04275	.99908	.04279	23.372	.0009	23.393	33
28	.04304	.99907	.04308	23.214	.0009	23.235	32
29	.04333	.99906	.04337	23.058	.0009	23.079	31
30	.04362	.99905	.04366	22.904	1.0009	22.925	30
31	.04391	.99903	.04395	22.752	.0010	22.774	29
32	.04420	.99902	.04424	22.602	.0010	22.624	28
33	.04449	.99901	.04453	22.454	.0010	22.476	27
34	.04478	.99900	.04483	22.308	.0010	22.330	26
35	.04507	.99898	.04512	22.164	1.0010	22.186	25
36	.04536	.99897	.04541	22.022	.0010	22.044	24
37	.04565	.99896	.04570	21.881	.0010	21.904	23
38	.04594	.99894	.04599	21.742	.0010	21.765	22
39	.04623	.99893	.04628	21.606	.0011	21.629	21
40	.04652	.99892	.04657	21.470	1.0011	21.494	20
41	.04681	.99890	.04687	21.337	.0011	21.360	19
42	.04711	.99889	.04716	21.205	.0011	21.228	18
43	.04740	.99888	.04745	21.075	.0011	21.098	17
44	.04769	.99886	.04774	20.946	.0011	20.970	16
45	.04798	.99885	.04803	20.819	1.0011	20.843	15
46	.04827	.99883	.04832	20.693	.0012	20.717	14
47	.04856	.99882	.04862	20.569	.0012	20.593	13
48	.04885	.99881	.04891	20.446	.0012	20.471	12
49	.04914	.99879	.04920	20.325	.0012	20.350	11
50	.04943	.99878	.04949	20.205	1.0012	20.230	10
51	.04972	.99876	.04978	20.087	.0012	20.112	9
52	.05001	.99875	.05007	19.970	.0012	19.995	8
53	.05030	.99873	.05037	19.854	.0013	19.880	7
54	.05059	.99872	.05066	19.740	.0013	19.766	6
55	.05088	.99870	.05095	19.627	1.0013	19.653	5
56	.05117	.99869	.05124	19.515	.0013	19.541	4
57	.05146	.99867	.05153	19.405	.0013	19.431	3
58	.05175	.99866	.05182	19.296	.0013	19.322	2
59	.05204	.99864	.05212	19.188	.0013	19.214	1
60	.05234	.99863	.05241	19.081	1.0014	19.107	0
M	Cosine	Sine	Cotan.	Tan.	Cosec.	Secant	M

87°

3°

M	Sine	Cosine	Tan.	Cotan.	Secant	Cosec.	M
0	.05234	.99863	.05241	19.081	1.0014	19.107	60
1	.05263	.99861	.05270	18.975	.0014	19.002	59
2	.05292	.99860	.05299	18.871	.0014	18.897	58
3	.05321	.99858	.05328	18.768	.0014	18.794	57
4	.05350	.99857	.05357	18.665	.0014	18.692	56
5	.05379	.99855	.05387	18.564	1.0014	18.591	55
6	.05408	.99854	.05416	18.464	.0015	18.491	54
7	.05437	.99852	.05445	18.365	.0015	18.393	53
8	.05466	.99850	.05474	18.268	.0015	18.295	52
9	.05495	.99849	.05503	18.171	.0015	18.198	51
10	.05524	.99847	.05532	18.075	1.0015	18.103	50
11	.05553	.99846	.05562	17.980	.0015	18.008	49
12	.05582	.99844	.05591	17.886	.0016	17.914	48
13	.05611	.99842	.05620	17.793	.0016	17.821	47
14	.05640	.99841	.05649	17.701	.0016	17.730	46
15	.05669	.99839	.05678	17.610	1.0016	17.639	45
16	.05698	.99837	.05707	17.520	.0016	17.549	44
17	.05727	.99836	.05737	17.431	.0016	17.460	43
18	.05756	.99834	.05766	17.343	.0017	17.372	42
19	.05785	.99832	.05795	17.256	.0017	17.285	41
20	.05814	.99831	.05824	17.169	1.0017	17.198	40
21	.05843	.99829	.05853	17.084	.0017	17.113	39
22	.05872	.99827	.05883	16.999	.0017	17.028	38
23	.05902	.99826	.05912	16.915	.0017	16.944	37
24	.05931	.99824	.05941	16.832	.0018	16.861	36
25	.05960	.99822	.05970	16.750	1.0018	16.779	35
26	.05989	.99820	.05999	16.668	.0018	16.698	34
27	.06018	.99819	.06029	16.587	.0018	16.617	33
28	.06047	.99817	.06058	16.507	.0018	16.538	32
29	.06076	.99815	.06087	16.428	.0018	16.459	31
30	.06105	.99813	.06116	16.350	1.0019	16.380	30
31	.06134	.99812	.06145	16.272	.0019	16.303	29
32	.06163	.99810	.06175	16.195	.0019	16.226	28
33	.06192	.99808	.06204	16.119	.0019	16.150	27
34	.06221	.99806	.06233	16.043	.0019	16.075	26
35	.06250	.99804	.06262	15.969	1.0019	16.000	25
36	.06279	.99803	.06291	15.894	.0020	15.926	24
37	.06308	.99801	.06321	15.821	.0020	15.853	23
38	.06337	.99799	.06350	15.748	.0020	15.780	22
39	.06366	.99797	.06379	15.676	.0020	15.708	21
40	.06395	.99795	.06408	15.605	1.0020	15.637	20
41	.06424	.99793	.06437	15.534	.0021	15.566	19
42	.06453	.99791	.06467	15.464	.0021	15.496	18
43	.06482	.99790	.06496	15.394	.0021	15.427	17
44	.06511	.99788	.06525	15.325	.0021	15.358	16
45	.06540	.99786	.06554	15.257	1.0021	15.290	15
46	.06569	.99784	.06583	15.189	.0022	15.222	14
47	.06598	.99782	.06613	15.122	.0022	15.155	13
48	.06627	.99780	.06642	15.056	.0022	15.089	12
49	.06656	.99778	.06671	14.990	.0022	15.023	11
50	.06685	.99776	.06700	14.924	1.0022	14.958	10
51	.06714	.99774	.06730	14.860	.0023	14.893	9
52	.06743	.99772	.06759	14.795	.0023	14.829	8
53	.06772	.99770	.06788	14.732	.0023	14.765	7
54	.06801	.99768	.06817	14.668	.0023	14.702	6
55	.06830	.99766	.06846	14.606	1.0023	14.640	5
56	.06859	.99764	.06876	14.544	.0024	14.578	4
57	.06888	.99762	.06905	14.482	.0024	14.517	3
58	.06918	.99760	.06934	14.421	.0024	14.456	2
59	.06947	.99758	.06963	14.361	.0024	14.395	1
60	.06976	.99756	.06993	14.301	1.0024	14.335	0
M	Cosine	Sine	Cotan.	Tan.	Cosec.	Secant	M

86°

4°

M	Sine	Cosine	Tan.	Cotan.	Secant	Cosec.	M
0	.06976	.99756	.06993	14.301	1.0024	14.335	60
1	.07005	.99754	.07022	14.241	.0025	14.276	59
2	.07034	.99752	.07051	14.182	.0025	14.217	58
3	.07063	.99750	.07080	14.123	.0025	14.159	57
4	.07092	.99748	.07110	14.065	.0025	14.101	56
5	.07121	.99746	.07139	14.008	1.0025	14.043	55
6	.07150	.99744	.07168	13.951	.0026	13.986	54
7	.07179	.99742	.07197	13.894	.0026	13.930	53
8	.07208	.99740	.07226	13.838	.0026	13.874	52
9	.07237	.99738	.07256	13.782	.0026	13.818	51
10	.07266	.99736	.07285	13.727	1.0026	13.763	50
11	.07295	.99733	.07314	13.672	.0027	13.708	49
12	.07324	.99731	.07343	13.617	.0027	13.654	48
13	.07353	.99729	.07373	13.563	.0027	13.600	47
14	.07382	.99727	.07402	13.510	.0027	13.547	46
15	.07411	.99725	.07431	13.457	1.0027	13.494	45
16	.07440	.99723	.07460	13.404	.0028	13.441	44
17	.07469	.99721	.07490	13.351	.0028	13.389	43
18	.07498	.99718	.07519	13.299	.0028	13.337	42
19	.07527	.99716	.07548	13.248	.0028	13.286	41
20	.07556	.99714	.07577	13.197	1.0029	13.235	40
21	.07585	.99712	.07607	13.146	.0029	13.184	39
22	.07614	.99710	.07636	13.096	.0029	13.134	38
23	.07643	.99707	.07665	13.046	.0029	13.084	37
24	.07672	.99705	.07694	12.996	.0029	13.034	36
25	.07701	.99703	.07724	12.947	1.0030	12.985	35
26	.07730	.99701	.07753	12.898	1.0030	12.937	34
27	.07759	.99698	.07782	12.849	.0030	12.888	33
28	.07788	.99696	.07812	12.801	.0030	12.840	32
29	.07817	.99694	.07841	12.754	.0031	12.793	31
30	.07846	.99692	.07870	12.706	1.0031	12.745	30
31	.07875	.99689	.07899	12.659	.0031	12.698	29
32	.07904	.99687	.07929	12.612	.0031	12.652	28
33	.07933	.99685	.07958	12.566	.0032	12.606	27
34	.07962	.99682	.07987	12.520	.0032	12.560	26
35	.07991	.99680	.08016	12.474	1.0032	12.514	25
36	.08020	.99678	.08046	12.429	.0032	12.469	24
37	.08049	.99675	.08075	12.384	.0032	12.424	23
38	.08078	.99673	.08104	12.339	.0033	12.379	22
39	.08107	.99671	.08134	12.295	.0033	12.335	21
40	.08136	.99668	.08163	12.250	1.0033	12.291	20
41	.08165	.99666	.08192	12.207	.0033	12.248	19
42	.08194	.99664	.08221	12.163	.0034	12.204	18
43	.08223	.99661	.08251	12.120	.0034	12.161	17
44	.08252	.99659	.08280	12.077	.0034	12.118	16
45	.08281	.99656	.08309	12.035	1.0034	12.076	15
46	.08310	.99654	.08339	11.992	.0035	12.034	14
47	.08339	.99652	.08368	11.950	.0035	11.992	13
48	.08368	.99649	.08397	11.909	.0035	11.950	12
49	.08397	.99647	.08426	11.867	.0035	11.909	11
50	.08426	.99644	.08456	11.826	1.0036	11.868	10
51	.08455	.99642	.08485	11.785	.0036	11.828	9
52	.08484	.99639	.08514	11.745	.0036	11.787	8
53	.08513	.99637	.08544	11.704	.0036	11.747	7
54	.08542	.99634	.08573	11.664	.0037	11.707	6
55	.08571	.99632	.08602	11.625	1.0037	11.668	5
56	.08600	.99629	.08632	11.585	.0037	11.628	4
57	.08629	.99627	.08661	11.546	1.0037	11.589	3
58	.08658	.99624	.08690	11.507	.0038	11.550	2
59	.08687	.99622	.08719	11.468	.0038	11.512	1
60	.08715	.99619	.08749	11.430	1.0038	11.474	0
M	Cosine	Sine	Cotan.	Tan.	Cosec.	Secant	M

85°

5°

M	Sine	Cosine	Tan.	Cotan.	Secant	Cosec.	M
0	.08715	.99619	.08749	11.430	1.0038	11.474	60
1	.08744	.99617	.08778	11.392	.0038	11.436	59
2	.08773	.99614	.08807	11.354	.0039	11.398	58
3	.08802	.99612	.08837	11.316	.0039	11.360	57
4	.08831	.99609	.08866	11.279	.0039	11.323	56
5	.08860	.99607	.08895	11.242	1.0039	11.286	55
6	.08889	.99604	.08925	11.205	.0040	11.249	54
7	.08918	.99601	.08954	11.168	.0040	11.213	53
8	.08947	.99599	.08983	11.132	.0046	11.176	52
9	.08976	.99596	.09013	11.095	.0040	11.140	51
10	.09005	.99594	.09042	11.059	1.0041	11.104	50
11	.09034	.99591	.09071	11.024	.0041	11.069	49
12	.09063	.99588	.09101	10.988	.0041	11.033	48
13	.09092	.99586	.09130	10.953	.0041	10.998	47
14	.09121	.99583	.09159	10.918	.0042	10.963	46
15	.09150	.99580	.09189	10.883	1.0042	10.929	45
16	.09179	.99578	.09218	10.848	.0042	10.894	44
17	.09208	.99575	.09247	10.814	.0043	10.860	43
18	.09237	.99572	.09277	10.780	.0043	10.826	42
19	.09266	.99570	.09306	10.746	.0043	10.792	41
20	.09295	.99567	.09335	10.712	1.0043	10.758	40
21	.09324	.99564	.09365	10.678	.0044	10.725	39
22	.09353	.99562	.09394	10.645	.0044	10.692	38
23	.09382	.99559	.09423	10.612	.0044	10.659	37
24	.09411	.99556	.09453	10.579	.0044	10.626	36
25	.09440	.99553	.09482	10.546	1.0045	10.593	35
26	.09469	.99551	.09511	10.514	.0045	10.561	34
27	.09498	.99548	.09541	10.481	.0045	10.529	33
28	.09527	.99545	.09570	10.449	.0046	10.497	32
29	.09556	.99542	.09599	10.417	.0046	10.465	31
30	.09584	.99540	.09629	10.385	1.0046	10.433	30
31	.09613	.99537	.09658	10.354	.0046	10.402	29
32	.09642	.99534	.09688	10.322	.0047	10.371	28
33	.09671	.99531	.09717	10.291	.0047	10.340	27
34	.09700	.99528	.09746	10.260	.0047	10.309	26
35	.09729	.99525	.09776	10.229	1.0048	10.278	25
36	.09758	.99523	.09805	10.199	.0048	10.248	24
37	.09787	.99520	.09834	10.168	.0048	10.217	23
38	.09816	.99517	.09864	10.138	.0048	10.187	22
39	.09845	.99514	.09893	10.108	.0049	10.157	21
40	.09874	.99511	.09922	10.078	1.0049	10.127	20
41	.09903	.99508	.09952	10.048	.0049	10.098	19
42	.09932	.99505	.09981	10.019	.0050	10.068	18
43	.09961	.99503	.10011	9.9893	.0050	10.039	17
44	.09990	.99500	.10040	9.9601	.0050	10.010	16
45	.10019	.99497	.10069	9.9310	1.0050	9.9812	15
46	.10048	.99494	.10099	9.9021	.0051	9.9525	14
47	.10077	.99491	.10128	9.8734	.0051	9.9239	13
48	.10106	.99488	.10158	9.8448	.0051	9.8955	12
49	.10134	.99485	.10187	9.8164	.0052	9.8672	11
50	.10163	.99482	.10216	9.7882	1.0052	9.8391	10
51	.10192	.99479	.10246	9.7601	.0052	9.8112	9
52	.10221	.99476	.10275	9.7322	.0053	9.7834	8
53	.10250	.99473	.10305	9.7044	.0053	9.7558	7
54	.10279	.99470	.10334	9.6768	.0053	9.7283	6
55	.10308	.99467	.10363	9.6493	1.0053	9.7010	5
56	.10337	.99464	.10393	9.6220	.0054	9.6739	4
57	.10366	.99461	.10422	9.5949	.0054	9.6469	3
58	.10395	.99458	.10452	9.5679	.0054	9.6200	2
59	.10424	.99455	.10481	9.5411	.0055	9.5933	1
60	.10453	.99452	.10510	9.5144	1.0055	9.5668	0
M	Cosine	Sine	Cotan.	Tan.	Cosec.	Secant	M

84°

M	Sine	Cosine	Tan.	Cotan.	Secant	Cosec.	M
0	.10453	.99452	.10510	9.5144	1.0055	9.5668	60
1	.10482	.99449	.10540	.4878	.0055	.5404	59
2	.10511	.99446	.10569	.4614	.0056	.5141	58
3	.10540	.99443	.10599	.4351	.0056	.4880	57
4	.10568	.99440	.10628	.4090	.0056	.4620	56
5	.10597	.99437	.10657	9.3831	1.0057	9.4362	55
6	.10626	.99434	.10687	.3572	.0057	.4105	54
7	.10655	.99431	.10716	.3315	.0057	.3850	53
8	.10684	.99428	.10746	.3060	.0057	.3596	52
9	.10713	.99424	.10775	.2806	.0058	.3343	51
10	.10742	.99421	.10805	9.2553	1.0058	9.3092	50
11	.10771	.99418	.10834	.2302	.0058	.2842	49
12	.10800	.99415	.10863	.2051	.0059	.2593	48
13	.10829	.99412	.10893	.1803	.0059	.2346	47
14	.10858	.99409	.10922	.1555	.0059	.2100	46
15	.10887	.99406	.10952	9.1309	1.0060	9.1855	45
16	.10916	.99402	.10981	.1064	.0060	.1612	44
17	.10944	.99399	.11011	.0821	.0060	.1370	43
18	.10973	.99396	.11040	.0579	.0061	.1129	42
19	.11002	.99393	.11069	.0338	0061	.0890	41
20	.11031	.99390	.11099	9.0098	1.0061	9.0651	40
21	.11060	.99386	.11128	8.9860	.0062	.0414	39
22	.11089	.99383	.11158	.9623	.0062	.0179	38
23	.11118	.99380	.11187	.9387	.0062	8.9944	37
24	.11147	.99377	.11217	.9152	.0063	.9711	36
25	.11176	.99373	.11246	8.8918	1.0063	8 9479	35
26	.11205	.99370	.11276	.8686	.0063	.9248	34
27	.11234	.99367	.11305	.8455	.0064	.9018	33
28	.11262	.99364	.11335	.8225	.0064	.8790	32
29	.11291	.99360	.11364	.7996	.0064	.8563	31
30	.11320	.99357	.11393	8.7769	1.0065	8.8337	30
31	.11349	.99354	.11423	.7542	.0065	.8112	29
32	.11378	.99350	.11452	.7317	.0065	.7888	28
33	.11407	.99347	.11482	.7093	.0066	.7665	27
34	.11436	.99344	.11511	.6870	.0066	.7444	26
35	.11465	.99341	.11541	8.6648	1.0066	8.7223	25
36	.11494	.99337	.11570	.6427	.0067	.7004	24
37	.11523	.99334	.11600	.6208	.0067	.6786	23
38	.11551	.99330	.11629	.5989	.0067	.6569	22
39	.11580	.99327	.11659	.5772	.0068	.6353	21
40	.11609	.99324	.11688	8.5555	1.0068	8.6138	20
41	.11638	.99320	.11718	.5340	.0068	.5924	19
42	.11667	.99317	.11747	.5126	.0069	.5711	18
43	.11696	.99314	.11777	.4913	.0069	.5499	17
44	.11725	.99310	.11806	.4701	.0069	.5289	16
45	.11754	.99307	.11836	8.4489	1.0070	8.5079	15
46	.11783	.99303	.11865	.4279	.0070	.4871	14
47	.11811	.99300	.11895	.4070	.0070	.4663	13
48	.11840	.99296	.11924	.3862	.0071	.4457	12
49	.11869	.99293	.11954	.3655	.0071	.4251	11
50	.11898	.99290	.11983	8.3449	1.0071	8.4046	10
51	.11927	.99286	.12013	.3244	.0072	.3843	9
52	.11956	.99283	.12042	.3040	.0072	.3640	8
53	.11985	.99279	.12072	.2837	.0073	.3439	7
54	.12014	.99276	.12101	.2635	.0073	.3238	6
55	.12042	.99272	.12131	8.2434	1.0073	8.3039	5
56	.12071	.99269	.12160	.2234	.0074	.2840	4
57	.12100	.99265	.12190	.2035	.0074	.2642	3
58	.12129	.99262	.12219	.1837	.0074	.2446	2
59	.12158	.99258	.12249	.1640	.0075	.2250	1
60	.12187	.99255	.12278	8.1443	1.0075	8.2055	0
M	Cosine	Sine	Cotan.	Tan.	Cosec.	Secant	M

83°

7°

M	Sine	Cosine	Tan.	Cotan.	Secant	Cosec.	M
0	.12187	.99255	.12278	8.1443	1.0075	8.2055	60
1	.12216	.99251	.12308	.1248	.0075	.1861	59
2	.12245	.99247	.12337	.1053	.0076	.1668	58
3	.12273	.99244	.12367	.0860	.0076	.1476	57
4	.12302	.99240	.12396	.0667	.0076	.1285	56
5	.12331	.99237	.12426	8.0476	1.0077	8.1094	55
6	.12360	.99233	.12456	.0285	.0077	.0905	54
7	.12389	.99229	.12485	.0095	.0078	.0717	53
8	.12418	.99226	.12515	7.9906	.0078	.0529	52
9	.12447	.99222	.12544	.9717	.0078	.0342	51
10	.12476	.99219	.12574	7.9530	1.0079	8.0156	50
11	.12504	.99215	.12603	.9344	.0079	7.9971	49
12	.12533	.99211	.12633	.9158	.0079	.9787	48
13	.12562	.99208	.12662	.8973	.0080	.9604	47
14	.12591	.99204	.12692	.8789	.0080	.9421	46
15	.12620	.99200	.12722	7.8606	1.0080	7.9240	45
16	.12649	.99197	.12751	.8424	.0081	.9059	44
17	.12678	.99193	.12781	.8243	.0081	.8879	43
18	.12706	.99189	.12810	.8062	.0082	.8700	42
19	.12735	.99186	.12840	.7882	.0082	.8522	41
20	.12764	.99182	.12869	7.7703	1.0082	7.8344	40
21	.12793	.99178	.12899	.7525	.0083	.8168	39
22	.12822	.99174	.12928	.7348	.0083	.7992	38
23	.12851	.99171	.12958	.7171	.0084	.7817	37
24	.12879	.99167	.12988	.6996	.0084	.7642	36
25	.12908	.99163	.13017	7.6821	1.0084	7.7469	35
26	.12937	.99160	.13047	.6646	.0085	.7296	34
27	.12966	.99156	.13076	.6473	.0085	.7124	33
28	.12995	.99152	.13106	.6300	.0085	.6953	32
29	.13024	.99148	.13136	.6129	.0086	.6783	31
30	.13053	.99144	.13165	7.5957	1.0086	7.6613	30
31	.13081	.99141	.13195	.5787	.0087	.6444	29
32	.13110	.99137	.13224	.5617	.0087	.6276	28
33	.13139	.99133	.13254	.5449	.0087	.6108	27
34	.13168	.99129	.13284	.5280	.0088	.5942	26
35	.13197	.99125	.13313	7.5113	1.0088	7.5776	25
36	.13226	.99121	.13343	.4946	.0089	.5611	24
37	.13254	.99118	.13372	.4780	.0089	.5446	23
38	.13283	.99114	.13402	.4615	.0089	.5282	22
39	.13312	.99110	.13432	.4451	.0090	.5119	21
40	.13341	.99106	.13461	7.4287	1.0090	7.4957	20
41	.13370	.99102	.13491	.4124	.0090	.4795	19
42	.13399	.99098	.13520	.3961	.0091	.4634	18
43	.13427	.99094	.13550	.3800	.0091	.4474	17
44	.13456	.99090	.13580	.3639	.0092	.4315	16
45	.13485	.99086	.13609	7.3479	1 0092	7.4156	15
46	.13514	.99083	.13639	.3319	.0092	.3998	14
47	.13543	.99079	.13669	.3160	.0093	.3840	13
48	.13571	.99075	.13698	.3002	.0093	.3683	12
49	.13600	.99071	.13728	.2844	.0094	.3527	11
50	.13629	.99067	.13757	7.2687	1.0094	7.3372	10
51	.13658	.99063	.13787	.2531	.0094	.3217	9
52	.13687	.99059	.13817	.2375	.0095	.3063	8
53	.13716	.99055	.13846	.2220	.0095	.2909	7
54	.13744	.99051	.13876	.2066	.0096	.2757	6
55	.13773	.99047	.13906	7.1912	1.0096	7.2604	5
56	.13802	.99043	.13935	.1759	.0097	.2453	4
57	.13831	.99039	.13965	.1607	.0097	.2302	3
58	.13860	.99035	.13995	.1455	.0097	.2152	2
59	.13888	.99031	.14024	.1304	.0098	.2002	1
60	.13917	.99027	.14054	7.1154	1.0098	7.1853	0
M	Cosine	Sine	Cotan.	Tan.	Cosec.	Secant	M

82°

8°

M	Sine	Cosine	Tan.	Cotan.	Secant	Cosec.	M
0	.13917	.99027	.14054	7.1154	1.0098	7.1853	60
1	.13946	.99023	.14084	.1004	.0099	.1704	59
2	.13975	.99019	.14113	.0854	.0099	.1557	58
3	.14004	.99015	.14143	.0706	.0099	.1409	57
4	.14032	.99010	.14173	.0558	.0100	.1263	56
5	.14061	.99006	.14202	7.0410	1.0100	7.1117	55
6	.14090	.99002	.14232	.0264	.0101	.0972	54
7	.14119	.98998	.14262	.0117	.0101	.0827	53
8	.14148	.98994	.14291	6.9972	.0102	.0683	52
9	.14176	.98990	.14321	.9827	.0102	.0539	51
10	.14205	.98986	.14351	6.9682	1.0102	7.0396	50
11	.14234	.98982	.14380	.9538	.0103	.0254	49
12	.14263	.98978	.14410	.9395	.0103	.0112	48
13	.14292	.98973	.14440	.9252	.0104	6.9971	47
14	.14320	.98969	.14470	.9110	.0104	.9830	46
15	.14349	.98965	.14499	6.8969	1.0104	6.9690	45
16	.14378	.98961	.14529	.8828	.0105	.9550	44
17	.14407	.98957	.14559	.8687	.0105	.9411	43
18	.14436	.98952	.14588	.8547	.0106	.9273	42
19	.14464	.98948	.14618	.8408	.0106	.9135	41
20	.14493	.98944	.14648	6.8269	1.0107	6.8998	40
21	.14522	.98940	.14677	.8131	.0107	.8861	39
22	.14551	.98936	.14707	.7993	.0107	.8725	38
23	.14579	.98931	.14737	.7856	.0108	.8589	37
24	.14608	.98927	.14767	.7720	.0108	.8454	36
25	.14637	.98923	.14796	6.7584	1.0109	6.8320	35
26	.14666	.98919	.14826	.7448	.0109	.8185	34
27	.14695	.98914	.14856	.7313	.0110	.8052	33
28	.14723	.98910	.14886	.7179	.0110	.7919	32
29	.14752	.98906	.14915	.7045	.0111	.7787	31
30	.14781	.98901	.14945	6.6911	1.0111	6.7655	30
31	.14810	.98897	.14975	.6779	.0111	.7523	29
32	.14838	.98893	.15004	.6646	.0112	.7392	28
33	.14867	.98889	.15034	.6514	.0112	.7262	27
34	.14896	.98884	.15064	.6383	.0113	.7132	26
35	.14925	.98880	.15094	6.6252	1.0113	6.7003	25
36	.14953	.98876	.15123	.6122	.0114	.6874	24
37	.14982	.98871	.15153	.5992	.0114	.6745	23
38	.15011	.98867	.15183	.5863	.0115	.6617	22
39	.15040	.98862	.15213	.5734	.0115	.6490	21
40	.15068	.98858	.15243	6.5605	1.0115	6.6363	20
41	.15097	.98854	.15272	.5478	.0116	.6237	19
42	.15126	.98849	.15302	.5350	.0116	.6111	18
43	.15155	.98845	.15332	.5223	.0117	.5985	17
44	.15183	.98840	.15362	.5097	.0117	.5860	16
45	.15212	.98836	.15391	6.4971	1.0118	6.5736	15
46	.15241	.98832	.15421	.4845	.0118	.5612	14
47	.15270	.98827	.15451	.4720	.0119	.5488	13
48	.15298	.98823	.15481	.4596	.0119	.5365	12
49	.15328	.98818	.15511	.4472	.0119	.5243	11
50	.15356	.98814	.15540	6.4348	1.0120	6.5121	10
51	.15385	.98809	.15570	.4225	.0120	.4999	9
52	.15413	.98805	.15600	.4103	.0121	.4878	8
53	.15442	.98800	.15630	.3980	.0121	.4757	7
54	.15471	.98796	.15659	.3859	.0122	.4637	6
55	.15500	.98791	.15689	6.3737	1.0122	6.4517	5
56	.15528	.98787	.15719	.3616	.0123	.4398	4
57	.15557	.98782	.15749	.3496	.0123	.4279	3
58	.15586	.98778	.15779	.3376	.0124	.4160	2
59	.15615	.98773	.15809	.3257	.0124	.4042	1
60	.15643	.98769	.15838	6.3137	1.0125	6.3924	0
M	Cosine	Sine	Cotan.	Tan.	Cosec.	Secant	M

81°

9°

M	Sine	Cosine	Tan.	Cotan.	Secant	Cosec.	M
0	.15643	.98769	.15838	6.3137	1.0125	6.3924	60
1	.15672	.98764	.15868	.3019	.0125	.3807	59
2	.15701	.98760	.15898	.2901	.0125	.3690	58
3	.15730	.98755	.15928	.2783	.0126	.3574	57
4	.15758	.98750	.15958	.2665	.0126	.3458	56
5	.15787	.98746	.15987	6.2548	1.0127	6.3343	55
6	.15816	.98741	.16017	.2432	.0127	.3228	54
7	.15844	.98737	.16047	.2316	.0128	.3113	53
8	.15873	.98732	.16077	.2200	.0128	.2999	52
9	.15902	.98727	.16107	.2085	.0129	.2885	51
10	.15931	.98723	.16137	6.1970	1.0129	6.2772	50
11	.15959	.98718	.16167	.1856	.0130	.2659	49
12	.15988	.98714	.16196	.1742	.0130	.2546	48
13	.16017	.98709	.16226	.1628	.0131	.2434	47
14	.16045	.98704	.16256	.1515	.0131	.2322	46
15	.16074	.98700	.16286	6.1402	1.0132	6.2211	45
16	.16103	.98695	.16316	.1290	.0132	.2100	44
17	.16132	.98690	.16346	.1178	.0133	.1990	43
18	.16160	.98685	.16376	.1066	.0133	.1880	42
19	.16189	.98681	.16405	.0955	.0134	.1770	41
20	.16218	.98676	.16435	6.0844	1.0134	6.1661	40
21	.16246	.98671	.16465	.0734	.0135	.1552	39
22	.16275	.98667	.16495	.0624	.0135	.1443	38
23	.16304	.98662	.16525	.0514	.0136	.1335	37
24	.16333	.98657	.16555	.0405	.0136	.1227	36
25	.16361	.98652	.16585	6.0296	1.0136	6.1120	35
26	.16390	.98648	.16615	.0188	.0137	.1013	34
27	.16419	.98643	.16644	.0080	.0137	.0906	33
28	.16447	.98638	.16674	5.9972	.0138	.0800	32
29	.16476	.98633	.16704	.9865	.0138	.0694	31
30	.16505	.98628	.16734	5.9758	1.0139	6.0588	30
31	.16533	98624	.16764	.9651	.0139	.0483	29
32	.16562	.98619	.16794	.9545	.0140	.0379	28
33	.16591	.98614	.16824	.9439	.0140	.0274	27
34	.16619	.98609	.16854	.9333	.0141	.0170	26
35	.16648	.98604	.16884	5.9228	1.0141	6.0066	25
36	.16677	.98600	.16914	.9123	.0142	5.9963	24
37	.16705	.98595	.16944	.9019	.0142	.9860	23
38	.16734	.98590	.16973	.8915	.0143	.9758	22
39	.16763	.98585	.17003	.8811	.0143	.9655	21
40	.16791	.98580	.17033	5.8708	1.0144	5 9554	20
41	.16820	.98575	.17063	.8605	.0144	.9452	19
42	.16849	.98570	.17093	.8502	.0145	.9351	18
43	.16878	.98565	.17123	.8400	.0145	.9250	17
44	.16906	.98560	.17153	.8298	.0146	.9150	16
45	.16935	.98556	.17183	5.8196	1.0146	5.9049	15
46	.16964	.98551	.17213	.8095	.0147	.8950	14
47	.16992	.98546	.17243	.7994	.0147	.8850	13
48	.17021	.98541	.17273	.7894	.0148	.8751	12
49	.17050	.98536	.17303	.7794	.0148	.8652	11
50	.17078	.98531	.17333	5.7694	1.0149	5.8554	10
51	.17107	.98526	.17363	.7594	.0150	.8456	9
52	.17136	.98521	.17393	.7495	.0150	.8358	8
53	.17164	.98516	.17423	.7396	.0151	.8261	7
54	.17193	.98511	.17453	.7297	.0151	.8163	6
55	.17221	.98506	.17483	5.7199	1.0152	5.8067	5
56	.17250	.98501	.17513	.7101	.0152	.7970	4
57	.17279	.98496	.17543	.7004	.0153	.7874	3
58	.17307	.98491	.17573	.6906	.0153	.7778	2
59	.17336	.98486	.17603	.6809	.0154	.7683	1
60	.17365	.98481	.17633	5.6713	1.0154	5.7588	0
M	Cosine	Sine	Cotan.	Tan.	Cosec.	Secant	M

80°

10°

M	Sine	Cosine	Tan.	Cotan.	Secant	Cosec.	M
0	.17365	.98481	.17633	5.6713	1.0154	5.7588	60
1	.17393	.98476	.17663	.6616	.0155	.7493	59
2	.17422	.98471	.17693	.6520	.0155	.7398	58
3	.17451	.98465	.17723	.6425	.0156	.7304	57
4	.17479	.98460	.17753	.6329	.0156	.7210	56
5	.17508	.98455	.17783	5.6234	1.0157	5.7117	55
6	.17537	.98450	.17813	.6140	.0157	.7023	54
7	.17565	.98445	.17843	.6045	.0158	.6930	53
8	.17594	.98440	.17873	.5951	.0158	.6838	52
9	.17622	.98435	.17903	.5857	.0159	.6745	51
10	.17651	.98430	.17933	5.5764	1.0159	5.6653	50
11	.17680	.98425	.17963	.5670	.0160	.6561	49
12	.17708	.98419	.17993	.5578	.0160	.6470	48
13	.17737	.98414	.18023	.5485	.0161	.6379	47
14	.17766	.98409	.18053	.5393	.0162	.6288	46
15	.17794	.98404	.18083	5.5301	1.0162	5.6197	45
16	.17823	.98399	.18113	.5209	.0163	.6107	44
17	.17852	.98394	.18143	.5117	.0163	.6017	43
18	.17880	.98388	.18173	.5026	.0164	.5928	42
19	.17909	.98383	.18203	.4936	.0164	.5838	41
20	.17937	.98378	.18233	5.4845	1.0165	5.5749	40
21	.17966	.98373	.18263	.4755	.0165	.5660	39
22	.17995	.98368	.18293	.4665	.0166	.5572	38
23	.18023	.98362	.18323	.4575	.0166	.5484	37
24	.18052	.98357	.18353	.4486	.0167	.5396	36
25	.18080	.98352	.18383	5.4396	1.0167	5.5308	35
26	.18109	.98347	.18413	.4308	.0168	.5221	34
27	.18138	.98341	.18444	.4219	.0169	.5134	33
28	.18166	.98336	.18474	.4131	.0169	.5047	32
29	.18195	.98331	.18504	.4043	.0170	.4960	31
30	.18223	.98325	.18534	5.3955	1.0170	5.4874	30
31	.18252	.98320	.18564	.3868	.0171	.4788	29
32	.18281	.98315	.18594	.3780	.0171	.4702	28
33	.18309	.98309	.18624	.3694	.0172	.4617	27
34	.18338	.98304	.18654	.3607	.0172	.4532	26
35	.18366	.98299	.18684	5.3521	1.0173	5.4447	25
36	.18395	.98293	.18714	.3434	.0174	.4362	24
37	.18424	.98288	.18745	.3349	.0174	.4278	23
38	.18452	.98283	.18775	.3263	.0175	.4194	22
39	.18481	.98277	.18805	.3178	.0175	.4110	21
40	.18509	.98272	.18835	5.3093	1.0176	5.4026	20
41	.18538	.98267	.18865	.3008	.0176	.3943	19
42	.18567	.98261	.18895	.2923	.0177	.3860	18
43	.18595	.98256	.18925	.2839	.0177	.3777	17
44	.18624	.98250	.18955	.2755	.0178	.3695	16
45	.18652	.98245	.18985	5.2671	1.0179	5.3612	15
46	.18681	.98240	.19016	.2588	.0179	.3530	14
47	.18709	.98234	.19046	.2505	.0180	.3449	13
48	.18738	.98229	.19076	.2422	.0180	.3367	12
49	.18767	.98223	.19106	.2339	.0181	.3286	11
50	.18795	.98218	.19136	5.2257	1.0181	5.3205	10
51	.18824	.98212	.19166	.2174	.0182	.3124	9
52	.18852	.98207	.19197	.2092	.0182	.3044	8
53	.18881	.98201	.19227	.2011	.0183	.2963	7
54	.18909	.98196	.19257	.1929	.0184	.2883	6
55	.18938	.98190	.19287	5.1848	1.0184	5.2803	5
56	.18967	.98185	.19317	.1767	.0185	.2724	4
57	.18995	.98179	.19347	.1686	.0185	.2645	3
58	.19024	.98174	.19378	.1606	.0186	.2566	2
59	.19052	.98168	.19408	.1525	.0186	.2487	1
60	.19081	.98163	.19438	5.1445	1.0187	5.2408	0
M	Cosine	Sine	Cotan.	Tan.	Cosec.	Secant	M

79°

11°

M	Sine	Cosine	Tan.	Cotan.	Secant	Cosec.	M
0	.19081	.98163	.19438	5.1445	1.0187	5.2408	60
1	.19109	.98157	.19468	.1366	.0188	.2330	59
2	.19138	.98152	.19498	.1286	.0188	.2252	58
3	.19166	.98146	.19529	.1207	.0189	.2174	57
4	.19195	.98140	.19559	.1128	.0189	.2097	56
5	.19224	.98135	.19589	5.1049	1.0190	5.2019	55
6	.19252	.98129	.19619	.0970	.0191	.1942	54
7	.19281	.98124	.19649	.0892	.0191	.1865	53
8	.19309	.98118	.19680	.0814	.0192	.1788	52
9	.19338	.98112	.19710	.0736	.0192	.1712	51
10	.19366	.98107	.19740	5.0658	1.0193	5.1636	50
11	.19395	.98101	.19770	.0581	.0193	.1560	49
12	.19423	.98095	.19800	.0504	.0194	.1484	48
13	.19452	.98090	.19831	.0427	.0195	.1409	47
14	.19480	.98084	.19861	.0350	.0195	.1333	46
15	.19509	.98078	.19891	5.0273	1.0196	5.1258	45
16	.19537	.98073	.19921	.0197	.0196	.1183	44
17	.19566	.98067	.19952	.0121	.0197	.1109	43
18	.19595	.98061	.19982	.0045	.0198	.1034	42
19	.19623	.98056	.20012	4.9969	.0198	.0960	41
20	.19652	.98050	.20042	4.9894	1.0199	5.0886	40
21	.19680	.98044	.20073	.9819	.0199	.0812	39
22	.19709	.98039	.20103	.9744	.0200	.0739	38
23	.19737	.98033	.20133	.9669	.0201	.0666	37
24	.19766	.98027	.20163	.9594	.0201	.0593	36
25	.19794	.98021	.20194	4.9520	1.0202	5.0520	35
26	.19823	.98016	.20224	.9446	.0202	.0447	34
27	.19851	.98010	.20254	.9372	.0203	.0375	33
28	.19880	.98004	.20285	.9298	.0204	.0302	32
29	.19908	.97998	.20315	.9225	.0204	.0230	31
30	.19937	.97992	.20345	4.9151	1.0205	5.0158	30
31	.19965	.97987	.20375	.9078	.0205	.0087	29
32	.19994	.97981	.20406	.9006	.0206	.0015	28
33	.20022	.97975	.20436	.8933	.0207	4.9944	27
34	.20051	.97969	.20466	.8860	.0207	.9873	26
35	.20079	.97963	.20497	4.8788	1.0208	4.9802	25
36	.20108	.97957	.20527	.8716	.0208	.9732	24
37	.20136	.97952	.20557	.8644	.0209	.9661	23
38	.20165	.97946	.20588	.8573	.0210	.9591	22
39	.20193	.97940	.20618	.8501	.0210	.9521	21
40	.20222	.97934	.20648	4.8430	1.0211	4.9452	20
41	.20250	.97928	.20679	.8359	.0211	.9382	19
42	.20279	.97922	.20709	.8288	.0212	.9313	18
43	.20307	.97916	.20739	.8217	.0213	.9243	17
44	.20336	.97910	.20770	.8147	.0213	.9175	16
45	.20364	.97904	.20800	4.8077	1.0214	4.9106	15
46	.20393	.97899	.20830	.8007	.0215	.9037	14
47	.20421	.97893	.20861	.7937	.0215	.8969	13
48	.20450	.97887	.20891	.7867	.0216	.8901	12
49	.20478	.97881	.20921	.7798	.0216	.8833	11
50	.20506	.97875	.20952	4.7728	1.0217	4.8765	10
51	.20535	.97869	.20982	.7659	.0218	.8697	9
52	.20563	.97863	.21012	.7591	.0218	.8630	8
53	.20592	.97857	.21043	.7522	.0219	.8563	7
54	.20620	.97851	.21073	.7453	.0220	.8496	6
55	.20649	.97845	.21104	4.7385	1.0220	4.8429	5
56	.20677	.97839	.21134	.7317	.0221	.8362	4
57	.20706	.97833	.21164	.7249	.0221	.8296	3
58	.20734	.97827	.21195	.7181	.0222	.8229	2
59	.20763	.97821	.21225	.7114	.0223	.8163	1
60	.20791	.97815	.21256	4.7046	1.0223	4.8097	0
M	Cosine	Sine	Cotan.	Tan.	Cosec.	Secant	M

78°

12°

M	Sine	Cosine	Tan.	Cotan.	Secant	Cosec.	M
0	.20791	.97815	.21256	4.7046	1.0223	4.8097	60
1	.20820	.97809	.21286	.6979	.0224	.8032	59
2	.20848	.97803	.21316	.6912	.0225	.7966	58
3	.20876	.97797	.21347	.6845	.0225	.7901	57
4	.20905	.97790	.21377	.6778	.0226	.7835	56
5	.20933	.97784	.21408	4.6712	1.0226	4.7770	55
6	.20962	.97778	.21438	.6646	.0227	.7706	54
7	.20990	.97772	.21468	.6580	.0228	.7641	53
8	.21019	.97766	.21499	.6514	.0228	.7576	52
9	.21047	.97760	.21529	.6448	.0229	.7512	51
10	.21076	.97754	.21560	4.6382	1.0230	4.7448	50
11	.21104	.97748	.21590	.6317	.0230	.7384	49
12	.21132	.97741	.21621	.6252	.0231	.7320	48
13	.21161	.97735	.21651	.6187	.0232	.7257	47
14	.21189	.97729	.21682	.6122	.0232	.7193	46
15	.21218	.97723	.21712	4.6057	1.0233	4.7130	45
16	.21246	.97717	.21742	.5993	.0234	.7067	44
17	.21275	.97711	.21773	.5928	.0234	.7004	43
18	.21303	.97704	.21803	.5864	.0235	.6942	42
19	.21331	.97698	.21834	.5800	.0235	.6879	41
20	.21360	.97692	.21864	4.5736	1.0236	4.6817	40
21	.21388	.97686	.21895	.5673	.0237	.6754	39
22	.21417	.97680	.21925	.5609	.0237	.6692	38
23	.21445	.97673	.21956	.5546	.0238	.6631	37
24	.21473	.97667	.21986	.5483	.0239	.6569	36
25	.21502	.97661	.22017	4.5420	1.0239	4.6507	35
26	.21530	.97655	.22047	.5357	.0240	.6446	34
27	.21559	.97648	.22078	.5294	.0241	.6385	33
28	.21587	.97642	.22108	.5232	.0241	.6324	32
29	.21615	.97636	.22139	.5169	.0242	.6263	31
30	.21644	.97630	.22169	4.5107	1.0243	4.6201	30
31	.21672	.97623	.22200	.5045	.0243	.6142	29
32	.21701	.97617	.22230	.4983	.0244	.6081	28
33	.21729	.97611	.22261	.4921	.0245	.6021	27
34	.21757	.97604	.22291	.4860	.0245	.5961	26
35	.21786	.97598	.22322	4.4799	1.0246	4.5901	25
36	.21814	.97592	.22353	.4737	.0247	.5841	24
37	.21843	.97585	.22383	.4676	.0247	.5782	23
38	.21871	.97579	.22414	.4615	.0248	.5722	22
39	.21899	.97573	.22444	.4555	.0249	.5663	21
40	.21928	.97566	.22475	4.4494	1.0249	4.5604	20
41	.21956	.97560	.22505	.4434	.0250	.5545	19
42	.21985	.97553	.22536	.4373	.0251	.5486	18
43	.22013	.97547	.22566	.4313	.0251	.5428	17
44	.22041	.97541	.22597	.4253	.0252	.5369	16
45	.22070	.97534	.22628	4.4194	1.0253	4.5311	15
46	.22098	.97528	.22658	.4134	.0253	.5253	14
47	.22126	.97521	.22689	.4074	.0254	.5195	13
48	.22155	.97515	.22719	.4015	.0255	.5137	12
49	.22183	.97508	.22750	.3956	.0255	.5079	11
50	.22211	.97502	.22781	4.3897	1.0256	4.5021	10
51	.22240	.97495	.22811	.3838	.0257	.4964	9
52	.22268	.97489	.22842	.3779	.0257	.4907	8
53	.22297	.97483	.22872	.3721	.0258	.4850	7
54	.22325	.97476	.22903	.3662	.0259	.4793	6
55	.22353	.97470	.22934	4.3604	1.0260	4.4736	5
56	.22382	.97463	.22964	.3546	.0260	.4679	4
57	.22410	.97457	.22995	.3488	.0261	.4623	3
58	.22438	.97450	.23025	.3430	.0262	.4566	2
59	.22467	.97443	.23056	.3372	.0262	.4510	1
60	.22495	.97437	.23087	4.3315	1.0263	4.4454	0
M	Cosine	Sine	Cotan.	Tan.	Cosec.	Secant	M

77°

13°

M	Sine	Cosine	Tan.	Cotan.	Secant	Cosec.	M
0	.22495	.97437	.23087	4.3315	1.0263	4.4454	60
1	.22523	.97430	.23117	.3257	.0264	.4398	59
2	.22552	.97424	.23148	.3200	.0264	.4342	58
3	.22580	.97417	.23179	.3143	.0265	.4287	57
4	.22608	.97411	.23209	.3086	.0266	.4231	56
5	.22637	.97404	.23240	4.3029	1.0266	4.4176	55
6	.22665	.97398	.23270	.2972	.0267	.4121	54
7	.22693	.97391	.23301	.2916	.0268	.4065	53
8	.22722	.97384	.23332	.2859	.0268	.4011	52
9	.22750	.97378	.23363	.2803	.0269	.3956	51
10	.22778	.97371	.23393	4.2747	1.0270	4.3901	50
11	.22807	.97364	.23424	.2691	.0271	.3847	49
12	.22835	.97358	.23455	.2635	.0271	.3792	48
13	.22863	.97351	.23485	.2579	.0272	.3738	47
14	.22892	.97344	.23516	.2524	.0273	.3684	46
15	.22920	.97338	.23547	4.2468	1.0273	4.3630	45
16	.22948	.97331	.23577	.2413	.0274	.3576	44
17	.22977	.97324	.23608	.2358	.0275	.3522	43
18	.23005	.97318	.23639	.2303	.0276	.3469	42
19	.23033	.97311	.23670	.2248	.0276	.3415	41
20	.23061	.97304	.23700	4.2193	1.0277	4.3362	40
21	.23090	.97298	.23731	.2139	.0278	.3309	39
22	.23118	.97291	.23762	.2084	.0278	.3256	38
23	.23146	.97284	.23793	.2030	.0279	.3203	37
24	.23175	.97277	.23823	.1976	.0280	.3150	36
25	.23203	.97271	.23854	4.1921	1.0280	4.3098	35
26	.23231	.97264	.23885	.1867	.0281	.3045	34
27	.23260	.97257	.23916	.1814	.0282	.2993	33
28	.23288	.97250	.23946	.1760	.0283	.2941	32
29	.23316	.97244	.23977	.1706	.0283	.2888	31
30	.23344	.97237	.24008	4.1653	1.0284	4.2836	30
31	.23373	.97230	.24039	.1600	.0285	.2785	29
32	.23401	.97223	.24069	.1546	.0285	.2733	28
33	.23429	.97216	.24100	.1493	.0286	.2681	27
34	.23458	.97210	.24131	.1440	.0287	.2630	26
35	.23486	.97203	.24162	4.1388	1.0288	4.2579	25
36	.23514	.97196	.24192	.1335	.0288	.2527	24
37	.23542	.97189	.24223	.1282	.0289	.2476	23
38	.23571	.97182	.24254	.1230	.0290	.2425	22
39	.23599	.97175	.24285	.1178	.0291	.2375	21
40	.23627	.97169	.24316	4.1126	1.0291	4.2324	20
41	.23655	.97162	.24346	.1073	.0292	.2273	19
42	.23684	.97155	.24377	.1022	.0293	.2223	18
43	.23712	.97148	.24408	.0970	.0293	.2173	17
44	.23740	.97141	.24439	.0918	.0294	.2122	16
45	.23768	.97134	.24470	4.0867	1.0295	4.2072	15
46	.23797	.97127	.24501	.0815	.0296	.2022	14
47	.23825	.97120	.24531	.0764	.0296	.1972	13
48	.23853	.97113	.24562	.0713	.0297	.1923	12
49	.23881	.97106	.24593	.0662	.0298	.1873	11
50	.23910	.97099	.24624	4.0611	1.0299	4.1824	10
51	.23938	.97092	.24655	.0560	.0299	.1774	9
52	.23966	.97086	.24686	.0509	.0300	.1725	8
53	.23994	.97079	.24717	.0458	.0301	.1676	7
54	.24023	.97072	.24747	.0408	.0302	.1627	6
55	.24051	.97065	.24778	4.0358	1.0302	4.1578	5
56	.24079	.97058	.24809	.0307	.0303	.1529	4
57	.24107	.97051	.24840	.0257	.0304	.1481	3
58	.24136	.97044	.24871	.0207	.0305	.1432	2
59	.24164	.97037	.24902	.0157	.0305	.1384	1
60	.24192	.97029	.24933	4.0108	1.0306	4.1336	0
M	Cosine	Sine	Cotan.	Tan.	Cosec.	Secant	M

76°

14°

M	Sine	Cosine	Tan.	Cotan.	Secant	Cosec.	M
0	.24192	.97029	.24933	4.0108	1.0306	4.1336	60
1	.24220	.97022	.24964	.0058	.0307	.1287	59
2	.24249	.97015	.24995	.0009	.0308	.1239	58
3	.24277	.97008	.25025	3.9959	.0308	.1191	57
4	.24305	.97001	.25056	.9910	.0309	.1144	56
5	.24333	.96994	.25087	3.9861	1.0310	4.1096	55
6	.24361	.96987	.25118	.9812	.0311	.1048	54
7	.24390	.96980	.25149	.9763	.0311	.1001	53
8	.24418	.96973	.25180	.9714	.0312	.0953	52
9	.24446	.96966	.25211	.9665	.0313	.0906	51
10	.24474	.96959	.25242	3.9616	1.0314	4.0859	50
11	.24502	.96952	.25273	.9568	.0314	.0812	49
12	.24531	.96944	.25304	.9520	.0315	.0765	48
13	.24559	.96937	.25335	.9471	.0316	.0718	47
14	.24587	.96930	.25366	.9423	.0317	.0672	46
15	.24615	.96923	.25397	3.9375	1.0317	4.0625	45
16	.24643	.96916	.25428	.9327	.0318	.0579	44
17	.24672	.96909	.25459	.9279	.0319	.0532	43
18	.24700	.96901	.25490	.9231	.0320	.0486	42
19	.24728	.96894	.25521	.9184	.0320	.0440	41
20	.24756	.96887	.25552	3.9136	1.0321	4.0394	40
21	.24784	.96880	.25583	.9089	.0322	.0348	39
22	.24813	.96873	.25614	.9042	.0323	.0302	38
23	.24841	.96865	.25645	.8994	.0323	.0256	37
24	.24869	.96858	.25676	.8947	.0324	.0211	36
25	.24897	.96851	.25707	3.8900	1.0325	4.0165	35
26	.24925	.96844	.25738	.8853	.0326	.0120	34
27	.24953	.96836	.25769	.8807	.0327	.0074	33
28	.24982	.96829	.25800	.8760	.0327	.0029	32
29	.25010	.96822	.25831	.8713	.0328	3.9984	31
30	.25038	.96815	.25862	3.8667	1.0329	3.9939	30
31	.25066	.96807	.25893	.8621	.0330	.9894	29
32	.25094	.96800	.25924	.8574	.0330	.9850	28
33	.25122	.96793	.25955	.8528	.0331	.9805	27
34	.25151	.96785	.25986	.8482	.0332	.9760	26
35	.25179	.96778	.26017	3.8436	1.0333	3.9716	25
36	.25207	.96771	.26048	.8390	.0334	.9672	24
37	.25235	.96763	.26079	.8345	.0334	.9627	23
38	.25263	.96756	.26110	.8299	.0335	.9583	22
39	.25291	.96749	.26141	.8254	.0336	.9539	21
40	.25319	.96741	.26172	3.8208	1.0337	3.9495	20
41	.25348	.96734	.26203	.8163	.0338	.9451	19
42	.25376	.96727	.26234	.8118	.0338	.9408	18
43	.25404	.96719	.26266	.8073	.0339	.9364	17
44	.25432	.96712	.26297	.8027	.0340	.9320	16
45	.25460	.96704	.26328	3.7983	1.0341	3.9277	15
46	.25488	.96697	.26359	.7938	.0341	.9234	14
47	.25516	.96690	.26390	.7893	.0342	.9190	13
48	.25544	.96682	.26421	.7848	.0343	.9147	12
49	.25573	.96675	.26452	.7804	.0344	.9104	11
50	.25601	.96667	.26483	3.7759	1.0345	3.9061	10
51	.25629	.96660	.26514	.7715	.0345	.9018	9
52	.25657	.96652	.26546	.7671	.0346	.8976	8
53	.25685	.96645	.26577	.7627	.0347	.8933	7
54	.25713	.96638	.26608	.7583	.0348	.8890	6
55	.25741	.96630	.26639	3.7539	1.0349	3.8848	5
56	.25769	.96623	.26670	.7495	.0349	.8805	4
57	.25798	.96615	.26701	.7451	.0350	.8763	3
58	.25826	.96608	.26732	.7407	.0351	.8721	2
59	.25854	.96600	.26764	.7364	.0352	.8679	1
60	.25882	.96592	.26795	3.7320	1.0353	3.8637	0
M	Cosine	Sine	Cotan.	Tan.	Cosec.	Secant	M

75°

15°

M	Sine	Cosine	Tan.	Cotan.	Secant	Cosec.	M
0	.25882	.96592	.26795	3.7320	1.0353	3.8637	60
1	.25910	.96585	.26826	.7277	.0353	.8595	59
2	.25938	.96577	.26857	.7234	.0354	.8553	58
3	.25966	.96570	.26888	.7191	.0355	.8512	57
4	.25994	.96562	.26920	.7147	.0356	.8470	56
5	.26022	.96555	.26951	3.7104	1.0357	3.8428	55
6	.26050	.96547	.26982	.7062	.0358	.8387	54
7	.26078	.96540	.27013	.7019	.0358	.8346	53
8	.26107	.96532	.27044	.6976	.0359	.8304	52
9	.26135	.96524	.27076	.6933	.0360	.8263	51
10	.26163	.96517	.27107	3.6891	1.0361	3.8222	50
11	.26191	.96509	.27138	.6848	.0362	.8181	49
12	.26219	.96502	.27169	.6806	.0362	.8140	48
13	.26247	.96494	.27201	.6764	.0363	.8100	47
14	.26275	.96486	.27232	.6722	.0364	.8059	46
15	.26303	.96479	.27263	3.6679	1.0365	3.8018	45
16	.26331	.96471	.27294	.6637	.0366	.7978	44
17	.26359	.96463	.27326	.6596	.0367	.7937	43
18	.26387	.96456	.27357	.6554	.0367	.7897	42
19	.26415	.96448	.27388	.6512	.0368	.7857	41
20	.26443	.96440	.27419	3.6470	1.0369	3.7816	40
21	.26471	.96433	.27451	.6429	.0370	.7776	39
22	.26499	.96425	.27482	.6387	.0371	.7736	38
23	.26527	.96417	.27513	.6346	.0371	.7697	37
24	.26556	.96409	.27544	.6305	.0372	.7657	36
25	.26584	.96402	.27576	3.6263	1.0373	3.7617	35
26	.26612	.96394	.27607	.6222	.0374	.7577	34
27	.26640	.96386	.27638	.6181	.0375	.7538	33
28	.26668	.96378	.27670	.6140	.0376	.7498	32
29	.26696	.96371	.27701	.6100	.0376	.7459	31
30	.26724	.96363	.27732	3.6059	1.0377	3.7420	30
31	.26752	.96355	.27764	.6018	.0378	.7380	29
32	.26780	.96347	.27795	.5977	.0379	.7341	28
33	.26808	.96340	.27826	.5937	.0380	.7302	27
34	.26836	.96332	.27858	.5896	.0381	.7263	26
35	.26864	.96324	.27889	3.5856	1.0382	3.7224	25
36	.26892	.96316	.27920	.5816	.0382	.7186	24
37	.26920	.96308	.27952	.5776	.0383	.7147	23
38	.26948	.96301	.27983	.5736	.0384	.7108	22
39	.26976	.96293	.28014	.5696	.0385	.7070	21
40	.27004	.96285	.28046	3.5656	1.0386	3.7031	20
41	.27032	.96277	.28077	.5616	.0387	.6993	19
42	.27060	.96269	.28109	.5576	.0387	.6955	18
43	.27088	.96261	.28140	.5536	.0388	.6917	17
44	.27116	.96253	.28171	.5497	.0389	.6878	16
45	.27144	.96245	.28203	3.5457	1.0390	3.6840	15
46	.27172	.96238	.28234	.5418	.0391	.6802	14
47	.27200	.96230	.28266	.5378	.0392	.6765	13
48	.27228	.96222	.28297	.5339	.0393	.6727	12
49	.27256	.96214	.28328	.5300	.0393	.6689	11
50	.27284	.96206	.28360	3.5261	1.0394	3.6651	10
51	.27312	.96198	.28391	.5222	.0395	.6614	9
52	.27340	.96190	.28423	.5183	.0396	.6576	8
53	.27368	.96182	.28454	.5144	1.0397	.6539	7
54	.27396	.96174	.28486	.5105	.0398	.6502	6
55	.27424	.96166	.28517	3.5066	1.0399	3.6464	5
56	.27452	.96158	.28549	.5028	.0399	.6427	4
57	.27480	.96150	.28580	.4989	.0400	.6390	3
58	.27508	.96142	.28611	.4951	.0401	.6353	2
59	.27536	.96134	.28643	.4912	.0402	.6316	1
60	.27564	.96126	.28674	3.4874	1.0403	3.6279	0
M	Cosine	Sine	Cotan.	Tan.	Cosec.	Secant	M

74°

16°

M	Sine	Cosine	Tan.	Cotan.	Secant	Cosec.	M
0	.27564	.96126	.28674	3.4874	1.0403	3.6279	60
1	.27592	.96118	.28706	.4836	.0404	.6243	59
2	.27620	.96110	.28737	.4798	.0405	.6206	58
3	.27648	.96102	.28769	.4760	.0406	.6169	57
4	.27675	.96094	.28800	.4722	.0406	.6133	56
5	.27703	.96086	.28832	3.4684	1.0407	3.6096	55
6	.27731	.96078	.28863	.4646	.0408	.6060	54
7	.27759	.96070	.28895	.4608	.0409	.6024	53
8	.27787	.96062	.28926	.4570	.0410	.5987	52
9	.27815	.96054	.28958	.4533	.0411	.5951	51
10	.27843	.96045	.28990	3.4495	1.0412	3.5915	50
11	.27871	.96037	.29021	.4458	.0413	.5879	49
12	.27899	.96029	.29053	.4420	.0413	.5843	48
13	.27927	.96021	.29084	.4383	.0414	.5807	47
14	.27955	.96013	.29116	.4346	.0415	.5772	46
15	.27983	.96005	.29147	3.4308	1.0416	3.5736	45
16	.28011	.95997	.29179	.4271	.0417	.5700	44
17	.28039	.95989	.29210	.4234	.0418	.5665	43
18	.28067	.95980	.29242	.4197	.0419	.5629	42
19	.28094	.95972	.29274	.4160	.0420	.5594	41
20	.28122	.95964	.29305	3.4124	1.0420	3.5559	40
21	.28150	.95956	.29337	.4087	.0421	.5523	39
22	.28178	.95948	.29368	.4050	.0422	.5488	38
23	.28206	.95940	.29400	.4014	.0423	.5453	37
24	.28234	.95931	.29432	.3977	.0424	.5418	36
25	.28262	.95923	.29463	3.3941	1.0425	3.5383	35
26	.28290	.95915	.29495	.3904	.0426	.5348	34
27	.28318	.95907	.29526	.3868	.0427	.5313	33
28	.28346	.95898	.29558	.3832	.0428	.5279	32
29	.28374	.95890	.29590	.3795	.0428	.5244	31
30	.28401	.95882	.29621	3.3759	1.0429	3.5209	30
31	.28429	.95874	.29653	.3723	.0430	.5175	29
32	.28457	.95865	.29685	.3687	.0431	.5140	28
33	.28485	.95857	.29716	.3651	.0432	.5106	27
34	.28513	.95849	.29748	.3616	.0433	.5072	26
35	.28541	.95840	.29780	3.3580	1.0434	3.5037	25
36	.28569	.95832	.29811	.3544	.0435	.5003	24
37	.28597	.95824	.29843	.3509	.0436	.4969	23
38	.28624	.95816	.29875	.3473	.0437	3.4935	22
39	.28652	.95807	.29906	.3438	.0438	.4901	21
40	.28680	.95799	.29938	3.3402	1.0438	3.4867	20
41	.28708	.95791	.29970	.3367	.0439	.4833	19
42	.28736	.95782	.30001	.3332	.0440	.4799	18
43	.28764	.95774	.30033	.3296	.0441	.4766	17
44	.28792	.95765	.30065	.3261	.0442	.4732	16
45	.28820	.95757	.30096	3.3226	1.0443	3.4698	15
46	.28847	.95749	.30128	.3191	.0444	.4665	14
47	.28875	.95740	.30160	.3156	.0445	.4632	13
48	.28903	.95732	.30192	.3121	.0446	.4598	12
49	.28931	.95723	.30223	.3087	.0447	.4565	11
50	.28959	.95715	.30255	3.3052	1.0448	3.4532	10
51	.28987	.95707	.30287	.3017	.0448	.4498	9
52	.29014	.95698	.30319	3.2983	.0449	.4465	8
53	.29042	.95690	.30350	.2948	.0450	.4432	7
54	.29070	.95681	.30382	.2914	.0451	.4399	6
55	.29098	.95673	.30414	3.2879	1.0452	3.4366	5
56	.29126	.95664	.30446	.2845	.0453	.4334	4
57	.29154	.95656	.30478	.2811	.0454	.4301	3
58	.29181	.95647	.30509	.2777	.0455	.4268	2
59	.29209	.95639	.30541	.2742	.0456	.4236	1
60	.29237	.95630	.30573	3.2708	1.0457	3.4203	0
M	Cosine	Sine	Cotan.	Tan.	Cosec.	Secant	M

73°

17°

M	Sine	Cosine	Tan.	Cotan.	Secant	Cosec.	M
0	.29237	.95630	.30573	3.2708	1.0457	3.4203	60
1	.29265	.95622	.30605	.2674	.0458	.4170	59
2	.29293	.95613	.30637	.2640	.0459	.4138	58
3	.29321	.95605	.30668	.2607	.0460	.4106	57
4	.29348	.95596	.30700	.2573	.0461	.4073	56
5	.29376	.95588	.30732	3.2539	1.0461	3.4041	55
6	.29404	.95579	.30764	.2505	.0462	.4009	54
7	.29432	.95571	.30796	.2472	.0463	.3977	53
8	.29460	.95562	.30828	.2438	.0464	.3945	52
9	.29487	.95554	.30859	.2405	.0465	.3913	51
10	.29515	.95545	.30891	3.2371	1.0466	3.3881	50
11	.29543	.95536	.30923	.2338	.0467	.3849	49
12	.29571	.95528	.30955	.2305	.0468	.3817	48
13	.29598	.95519	.30987	.2271	.0469	.3785	47
14	.29626	.95511	.31019	.2238	.0470	.3754	46
15	.29654	.95502	.31051	3.2205	1.0471	3.3722	45
16	.29682	.95493	.31083	.2172	.0472	.3690	44
17	.29710	.95485	.31115	.2139	.0473	.3659	43
18	.29737	.95476	.31146	.2106	.0474	.3627	42
19	.29765	.95467	.31178	.2073	.0475	.3596	41
20	.29793	.95459	.31210	3.2041	1.0476	3.3565	40
21	.29821	.95450	.31242	.2008	.0477	.3534	39
22	.29848	.95441	.31274	.1975	.0478	.3502	38
23	.29876	.95433	.31306	.1942	.0478	.3471	37
24	.29904	.95424	.31338	.1910	.0479	.3440	36
25	.29932	.95415	.31370	3.1877	1.0480	3.3409	35
26	.29959	.95407	.31402	.1845	.0481	.3378	34
27	.29987	.95398	.31434	.1813	.0482	.3347	33
28	.30015	.95389	.31466	.1780	.0483	.3316	32
29	.30043	.95380	.31498	.1748	.0484	.3286	31
30	.30070	.95372	.31530	3.1716	1.0485	3.3255	30
31	.30098	.95363	.31562	.1684	.0486	.3224	29
32	.30126	.95354	.31594	.1652	.0487	.3194	28
33	.30154	.95345	.31626	.1620	.0488	.3163	27
34	.30181	.95337	.31658	.1588	.0489	.3133	26
35	.30209	.95328	.31690	3.1556	1.0490	3.3102	25
36	.30237	.95319	.31722	.1524	.0491	.3072	24
37	.30265	.95310	.31754	.1492	.0492	.3042	23
38	.30292	.95301	.31786	.1460	.0493	.3011	22
39	.30320	.95293	.31818	.1429	.0494	.2981	21
40	.30348	.95284	.31850	3.1397	1.0495	3.2951	20
41	.30375	.95275	.31882	.1366	.0496	.2921	19
42	.30403	.95266	.31914	.1334	.0497	.2891	18
43	.30431	.95257	.31946	.1303	.0498	.2861	17
44	.30459	.95248	.31978	.1271	.0499	.2831	16
45	.30486	.95239	.32010	3.1240	1.0500	3.2801	15
46	.30514	.95231	.32042	.1209	.0501	.2772	14
47	.30542	.95222	.32074	.1177	.0502	.2742	13
48	.30569	.95213	.32106	.1146	.0503	.2712	12
49	.30597	.95204	.32138	.1115	.0504	.2683	11
50	.30625	.95195	.32171	3.1084	1.0505	3.2653	10
51	.30653	.95186	.32203	.1053	.0506	.2624	9
52	.30680	.95177	.32235	.1022	.0507	.2594	8
53	.30708	.95168	.32267	.0991	.0508	.2565	7
54	.30736	.95159	.32299	.0960	.0509	.2535	6
55	.30763	.95150	.32331	3.0930	1.0510	3.2506	5
56	.30791	.95141	.32363	.0899	.0511	.2477	4
57	.30819	.95132	.32395	.0868	.0512	.2448	3
58	.30846	.95124	.32428	.0838	.0513	.2419	2
59	.30874	.95115	.32460	.0807	.0514	.2390	1
60	.30902	.95106	.32492	3.0777	1.0515	3.2361	0
M	Cosine	Sine	Cotan.	Tan.	Cosec.	Secant	M

72°

18°

M	Sine	Cosine	Tan.	Cotan.	Secant	Cosec.	M
0	.30902	.95106	.32492	3.0777	1.0515	3.2361	60
1	.30929	.95097	.32524	.0746	.0516	.2332	59
2	.30957	.95088	.32556	.0716	.0517	.2303	58
3	.30985	.95079	.32588	.0686	.0518	.2274	57
4	.31012	.95070	.32621	.0655	.0519	.2245	56
5	.31040	.95061	.32653	3.0625	1.0520	3.2216	55
6	.31068	.95051	.32685	.0595	.0521	.2188	54
7	.31095	.95042	.32717	.0565	.0522	.2159	53
8	.31123	.95033	.32749	.0535	.0523	.2131	52
9	.31150	.95024	.32782	.0505	.0524	.2102	51
10	.31178	.95015	.32814	3.0475	1.0525	3.2074	50
11	.31206	.95006	.32846	.0445	.0526	.2045	49
12	.31233	.94997	.32878	.0415	.0527	.2017	48
13	.31261	.94988	.32910	.0385	.0528	.1989	47
14	.31289	.94979	.32943	.0356	.0529	.1960	46
15	.31316	.94970	.32975	3.0326	1.0530	3.1932	45
16	.31344	.94961	.33007	.0296	.0531	.1904	44
17	.31372	.94952	.33039	.0267	.0532	.1876	43
18	.31399	.94942	.33072	.0237	.0533	.1848	42
19	.31427	.94933	.33104	.0208	.0534	.1820	41
20	.31454	.94924	.33136	3.0178	1.0535	3.1792	40
21	.31482	.94915	.33169	.0149	.0536	.1764	39
22	.31510	.94906	.33201	.0120	.0537	.1736	38
23	.31537	.94897	.33233	.0090	.0538	.1708	37
24	.31565	.94888	.33265	.0061	.0539	.1681	36
25	.31592	.94878	.33298	3.0032	1.0540	3.1653	35
26	.31620	.94869	.33330	.0003	.0541	.1625	34
27	.31648	.94860	.33362	2.9974	.0542	.1598	33
28	.31675	.94851	.33395	.9945	.0543	.1570	32
29	.31703	.94841	.33427	.9916	.0544	.1543	31
30	.31730	.94832	.33459	2.9887	1.0545	3.1515	30
31	.31758	.94823	.33492	.9858	.0546	.1488	29
32	.31786	.94814	.33524	.9829	.0547	.1461	28
33	.31813	.94805	.33557	.9800	.0548	.1433	27
34	.31841	.94795	.33589	.9772	.0549	.1406	26
35	.31868	.94786	.33621	2.9743	1.0550	3.1379	25
36	.31896	.94777	.33654	.9714	.0551	.1352	24
37	.31923	.94767	.33686	.9686	.0552	.1325	23
38	.31951	.94758	.33718	.9657	.0553	.1298	22
39	.31978	.94749	.33751	.9629	.0554	.1271	21
40	.32006	.94740	.33783	2.9600	1.0555	3.1244	20
41	.32034	.94730	.33816	.9572	.0556	.1217	19
42	.32061	.94721	.33848	.9544	.0557	.1190	18
43	.32089	.94712	.33880	.9515	.0558	.1163	17
44	.32116	.94702	.33913	.9487	.0559	.1137	16
45	.32144	.94693	.33945	2.9459	1.0560	3.1110	15
46	.32171	.94684	.33978	.9431	.0561	.1083	14
47	.32199	.94674	.34010	.9403	.0562	.1057	13
48	.32226	.94665	.34043	.9375	.0563	.1030	12
49	.32254	.94655	.34075	.9347	.0565	.1004	11
50	.32282	.94646	.34108	2.9319	1.0566	3.0977	10
51	.32309	.94637	.34140	.9291	.0567	.0951	9
52	.32337	.94627	.34173	.9263	.0568	.0925	8
53	.32364	.94618	.34205	.9235	.0569	.0898	7
54	.32392	.94608	.34238	.9208	.0570	.0872	6
55	.32419	.94599	.34270	2.9180	1.0571	3.0846	5
56	.32447	.94590	.34303	.9152	.0572	.0820	4
57	.32474	.94580	.34335	.9125	.0573	.0793	3
58	.32502	.94571	.34368	.9097	.0574	.0767	2
59	.32529	.94561	.34400	.9069	.0575	.0741	1
60	.32557	.94552	.34433	2.9042	1.0576	3.0715	0
M	**Cosine**	**Sine**	**Cotan.**	**Tan.**	**Cosec.**	**Secant**	**M**

71°

19°

M	Sine	Cosine	Tan.	Cotan.	Secant	Cosec.	M
0	.32557	.94552	.34433	2.9042	1.0576	3.0715	60
1	.32584	.94542	.34465	.9015	.0577	.0690	59
2	.32612	.94533	.34498	.8987	.0578	.0664	58
3	.32639	.94523	.34530	.8960	.0579	.0638	57
4	.32667	.94514	.34563	.8933	.0580	.0612	56
5	.32694	.94504	.34595	2.8905	1.0581	3.0586	55
6	.32722	.94495	.34628	.8878	.0582	.0561	54
7	.32749	.94485	.34661	.8851	.0584	.0535	53
8	.32777	.94476	.34693	.8824	.0585	.0509	52
9	.32804	.94466	.34726	.8797	.0586	.0484	51
10	.32832	.94457	.34758	2.8770	1.0587	3.0458	50
11	.32859	.94447	.34791	.8743	.0588	.0433	49
12	.32887	.94438	.34824	.8716	.0589	.0407	48
13	.32914	.94428	.34856	.8689	.0590	.0382	47
14	.32942	.94418	.34889	.8662	.0591	.0357	46
15	.32969	.94409	.34921	2.8636	1.0592	3.0331	45
16	.32996	.94399	.34954	.8609	.0593	.0306	44
17	.33024	.94390	.34987	.8582	.0594	.0281	43
18	.33051	.94380	.35019	.8555	.0595	.0256	42
19	.33079	.94370	.35052	.8529	.0596	.0231	41
20	.33106	.94361	.35085	2.8502	1.0598	3.0206	40
21	.33134	.94351	.35117	.8476	.0599	.0181	39
22	.33161	.94341	.35150	.8449	.0600	.0156	38
23	.33189	.94332	.35183	.8423	.0601	.0131	37
24	.33216	.94322	.35215	.8396	.0602	.0106	36
25	.33243	.94313	.35248	2.8370	1.0603	3.0081	35
26	.33271	.94303	.35281	.8344	.0604	.0056	34
27	.33298	.94293	.35314	.8318	.0605	.0031	33
28	.33326	.94283	.35346	.8291	.0606	.0007	32
29	.33353	.94274	.35379	.8265	.0607	2.9982	31
30	.33381	.94264	.35412	2.8239	1.0608	2.9957	30
31	.33408	.94254	.35445	.8213	.0609	.9933	29
32	.33435	.94245	.35477	.8187	.0611	.9908	28
33	.33463	.94235	.35510	.8161	.0612	.9884	27
34	.33490	.94225	.35543	.8135	.0613	.9859	26
35	.33518	.94215	.35576	2.8109	1.0614	2.9835	25
36	.33545	.94206	.35608	.8083	.0615	.9810	24
37	.33572	.94196	.35641	.8057	.0616	.9786	23
38	.33600	.94186	.35674	.8032	.0617	.9762	22
39	.33627	.94176	.35707	.8006	.0618	.9738	21
40	.33655	.94167	.35739	2.7980	1.0619	2.9713	20
41	.33682	.94157	.35772	.7954	.0620	.9689	19
42	.33709	.94147	.35805	.7929	.0622	.9665	18
43	.33737	.94137	.35838	.7903	.0623	.9641	17
44	.33764	.94127	.35871	.7878	.0624	.9617	16
45	.33792	.94118	.35904	2.7852	1.0625	2.9593	15
46	.33819	.94108	.35936	.7827	.0626	.9569	14
47	.33846	.94098	.35969	.7801	.0627	.9545	13
48	.33874	.94088	.36002	.7776	.0628	.9521	12
49	.33901	.94078	.36035	.7751	.0629	.9497	11
50	.33928	.94068	.36068	2.7725	1.0630	2.9474	10
51	.33956	.94058	.36101	.7700	.0632	.9450	9
52	.33983	.94049	.36134	.7675	.0633	.9426	8
53	.34011	.94039	.36167	.7650	.0634	.9402	7
54	.34038	.94029	.36199	.7625	.0635	.9379	6
55	.34065	.94019	.36232	2.7600	1.0636	2.9355	5
56	.34093	.94009	.36265	.7575	.0637	.9332	4
57	.34120	.93999	.36298	.7550	.0638	.9308	3
58	.34147	.93989	.36331	.7525	.0639	.9285	2
59	.34175	.93979	.36364	.7500	.0641	.9261	1
60	.34202	.93969	.36397	2.7475	1.0642	2.9238	0
M	Cosine	Sine	Cotan.	Tan.	Cosec.	Secant	M

70°

20°

M	Sine	Cosine	Tan.	Cotan.	Secant	Cosec.	M
0	.34202	.93969	.36397	2.7475	1.0642	2.9238	60
1	.34229	.93959	.36430	.7450	.0643	.9215	59
2	.34257	.93949	.36463	.7425	.0644	.9191	58
3	.34284	.93939	.36496	.7400	.0645	.9168	57
4	.34311	.93929	.36529	.7376	.0646	.9145	56
5	.34339	.93919	.36562	2.7351	1.0647	2.9122	55
6	.34366	.93909	.36595	.7326	.0648	.9098	54
7	.34393	.93899	.36628	.7302	.0650	.9075	53
8	.34421	.93889	.36661	.7277	.0651	.9052	52
9	.34448	.93879	.36694	.7252	.0652	.9029	51
10	.34475	.93869	.36727	2.7228	1.0653	2.9006	50
11	.34502	.93859	.36760	.7204	.0654	.8983	49
12	.34530	.93849	.36793	.7179	.0655	.8960	48
13	.34557	.93839	.36826	.7155	.0656	.8937	47
14	.34584	.93829	.36859	.7130	.0658	.8915	46
15	.34612	.93819	.36892	2.7106	1.0659	2.8892	45
16	.34639	.93809	.36925	.7082	.0660	.8869	44
17	.34666	.93799	.36958	.7058	.0661	.8846	43
18	.34693	.93789	.36991	.7033	.0662	.8824	42
19	.34721	.93779	.37024	.7009	.0663	.8801	41
20	.34748	.93769	.37057	2.6985	1.0664	2.8778	40
21	.34775	.93758	.37090	.6961	.0666	.8756	39
22	.34803	.93748	.37123	.6937	.0667	.8733	38
23	.34830	.93738	.37156	.6913	.0668	.8711	37
24	.34857	.93728	.37190	.6889	.0669	.8688	36
25	.34884	.93718	.37223	2.6865	1.0670	2.8666	35
26	.34912	.93708	.37256	.6841	.0671	.8644	34
27	.34939	.93698	.37289	.6817	.0673	.8621	33
28	.34966	.93687	.37322	.6794	.0674	.8599	32
29	.34993	.93677	.37355	.6770	.0675	.8577	31
30	.35021	.93667	.37388	2.6746	1.0676	2.8554	30
31	.35048	.93657	.37422	.6722	.0677	.8532	29
32	.35075	.93647	.37455	.6699	.0678	.8510	28
33	.35102	.93637	.37488	.6675	.0679	.8488	27
34	.35130	.93626	.37521	.6652	.0681	.8466	26
35	.35157	.93616	.37554	2.6628	1.0682	2.8444	25
36	.35184	.93606	.37587	.6604	.0683	.8422	24
37	.35211	.93596	.37621	.6581	.0684	.8400	23
38	.35239	.93585	.37654	.6558	.0685	.8378	22
39	.35266	.93575	.37687	.6534	.0686	.8356	21
40	.35293	.93565	.37720	2.6511	1.0688	2.8334	20
41	.35320	.93555	.37754	.6487	.0689	.8312	19
42	.35347	.93544	.37787	.6464	.0690	.8290	18
43	.35375	.93534	.37820	.6441	.0691	.8269	17
44	.35402	.93524	.37853	.6418	.0692	.8247	16
45	.35429	.93513	.37887	2.6394	1.0694	2.8225	15
46	.35456	.93503	.37920	.6371	.0695	.8204	14
47	.35483	.93493	.37953	.6348	.0696	.8182	13
48	.35511	.93482	.37986	.6325	.0697	.8160	12
49	.35538	.93472	.38020	.6302	.0698	.8139	11
50	.35565	.93462	.38053	2.6279	1.0699	2.8117	10
51	.35592	.93451	.38086	.6256	.0701	.8096	9
52	.35619	.93441	.38120	.6233	.0702	.8074	8
53	.35647	.93431	.38153	.6210	.0703	.8053	7
54	.35674	.93420	.38186	.6187	.0704	.8032	6
55	.35701	.93410	.38220	2.6164	1.0705	2.8010	5
56	.35728	.93400	.38253	.6142	.0707	.7989	4
57	.35755	.93389	.38286	.6119	.0708	.7968	3
58	.35782	.93379	.38320	.6096	.0709	.7947	2
59	.35810	.93368	.38353	.6073	.0710	.7925	1
60	.35837	.93358	.38386	2.6051	1.0711	2.7904	0
M	Cosine	Sine	Cotan.	Tan.	Cosec.	Secant	M

69°

21°

M	Sine	Cosine	Tan.	Cotan.	Secant	Cosec.	M
0	.35837	.93358	.38386	2.6051	1.0711	2.7904	60
1	.35864	.93348	.38420	.6028	.0713	.7883	59
2	.35891	.93337	.38453	.6006	.0714	.7862	58
3	.35918	.93327	.38486	.5983	.0715	.7841	57
4	.35945	.93316	.38520	.5960	.0716	.7820	56
5	.35972	.93306	.38553	2.5938	1.0717	2.7799	55
6	.36000	.93295	.38587	.5916	.0719	.7778	54
7	.36027	.93285	.38620	.5893	.0720	.7757	53
8	.36054	.93274	.38654	.5871	.0721	.7736	52
9	.36081	.93264	.38687	.5848	.0722	.7715	51
10	.36108	.93253	.38720	2.5826	1.0723	2.7694	50
11	.36135	.93243	.38754	.5804	.0725	.7674	49
12	.36162	.93232	.38787	.5781	.0726	.7653	48
13	.36189	.93222	.38821	.5759	.0727	.7632	47
14	.36217	.93211	.38854	.5737	.0728	.7611	46
15	.36244	.93201	.38888	2.5715	1.0729	2.7591	45
16	.36271	.93190	.38921	.5693	.0731	.7570	44
17	.36298	.93180	.38955	.5671	.0732	.7550	43
18	.36325	.93169	.38988	.5649	.0733	.7529	42
19	.36352	.93158	.39022	.5627	.0734	.7509	41
20	.36379	.93148	.39055	2.5605	1.0736	2.7488	40
21	.36406	.93137	.39089	.5583	.0737	.7468	39
22	.36433	.93127	.39122	.5561	.0738	.7447	38
23	.36460	.93116	.39156	.5539	.0739	.7427	37
24	.36488	.93105	.39189	.5517	.0740	.7406	36
25	.36515	.93095	.39223	2.5495	1.0742	2.7386	35
26	.36542	.93084	.39257	.5473	.0743	.7366	34
27	.36569	.93074	.39290	.5451	.0744	.7346	33
28	.36596	.93063	.39324	.5430	.0745	.7325	32
29	.36623	.93052	.39357	.5408	.0747	.7305	31
30	.36650	.93042	.39391	2.5386	1.0748	2.7285	30
31	.36677	.93031	.39425	.5365	.0749	.7265	29
32	.36704	.93020	.39458	.5343	.0750	.7245	28
33	.36731	.93010	.39492	.5322	.0751	.7225	27
34	.36758	.92999	.39525	.5300	.0753	.7205	26
35	.36785	.92988	.39559	2.5278	1.0754	2.7185	25
36	.36812	.92978	.39593	.5257	.0755	.7165	24
37	.36839	.92967	.39626	.5236	.0756	.7145	23
38	.36866	.92956	.39660	.5214	.0758	.7125	22
39	.36893	.92945	.39694	.5193	.0759	.7105	21
40	.36921	.92935	.39727	2.5171	1.0760	2.7085	20
41	.36948	.92924	.39761	.5150	.0761	.7065	19
42	.36975	.92913	.39795	.5129	.0763	.7045	18
43	.37002	.92902	.39828	.5108	.0764	.7026	17
44	.37029	.92892	.39862	.5086	.0765	.7006	16
45	.37056	.92881	.39896	2.5065	1.0766	2.6986	15
46	.37083	.92870	.39930	.5044	.0768	.6967	14
47	.37110	.92859	.39963	.5023	.0769	.6947	13
48	.37137	.92848	.39997	.5002	.0770	.6927	12
49	.37164	.92838	.40031	.4981	.0771	.6908	11
50	.37191	.92827	.40065	2.4960	1.0773	2.6888	10
51	.37218	.92816	.40098	.4939	.0774	.6869	9
52	.37245	.92805	.40132	.4918	.0775	.6849	8
53	.37272	.92794	.40166	.4897	.0776	.6830	7
54	.37299	.92784	.40200	.4876	.0778	.6810	6
55	.37326	.92773	.40233	2.4855	1.0779	2.6791	5
56	.37353	.92762	.40267	.4834	.0780	.6772	4
57	.37380	.92751	.40301	.4813	.0781	.6752	3
58	.37407	.92740	.40335	.4792	.0783	.6733	2
59	.37434	.92729	.40369	.4772	.0784	.6714	1
60	.37461	.92718	.40403	2.4751	1.0785	2.6695	0
M	Cosine	Sine	Cotan.	Tan.	Cosec.	Secant	M

68°

22°

M	Sine	Cosine	Tan.	Cotan.	Secant	Cosec.	M
0	.37461	.92718	.40403	2.4751	1.0785	2.6695	60
1	.37488	.92707	.40436	.4730	.0787	.6675	59
2	.37514	.92696	.40470	.4709	.0788	.6656	58
3	.37541	.92686	.40504	.4689	.0789	.6637	57
4	.37568	.92675	.40538	.4668	.0790	.6618	56
5	.37595	.92664	.40572	2.4647	1.0792	2.6599	55
6	.37622	.92653	.40606	.4627	.0793	.6580	54
7	.37649	.92642	.40640	.4606	.0794	.6561	53
8	.37676	.92631	.40673	.4586	.0795	.6542	52
9	.37703	.92620	.40707	.4565	.0797	.6523	51
10	.37730	.92609	.40741	2.4545	1.0798	2.6504	50
11	.37757	.92598	.40775	.4525	.0799	.6485	49
12	.37784	.92587	.40809	.4504	.0801	.6466	48
13	.37811	.92576	.40843	.4484	.0802	.6447	47
14	.37838	.92565	.40877	.4463	.0803	.6428	46
15	.37865	.92554	.40911	2.4443	1.0804	2.6410	45
16	.37892	.92543	.40945	.4423	.0806	.6391	44
17	.37919	.92532	.40979	.4403	.0807	.6372	43
18	.37946	.92521	.41013	.4382	.0808	.6353	42
19	.37972	.92510	.41047	.4362	.0810	.6335	41
20	.37999	.92499	.41081	2.4342	1.0811	2.6316	40
21	.38026	.92488	.41115	.4322	.0812	.6297	39
22	.38053	.92477	.41149	.4302	.0813	.6279	38
23	.38080	.92466	.41183	.4282	.0815	.6260	37
24	.38107	.92455	.41217	.4262	.0816	.6242	36
25	.38134	.92443	.41251	2.4242	1.0817	2.6223	35
26	.38161	.92432	.41285	.4222	.0819	.6205	34
27	.38188	.92421	.41319	.4202	.0820	.6186	33
28	.38214	.92410	.41353	.4182	.0821	.6168	32
29	.38241	.92399	.41387	.4162	.0823	.6150	31
30	.38268	.92388	.41421	2.4142	1.0824	2.6131	30
31	.38295	.92377	.41455	.4122	.0825	.6113	29
32	.38322	.92366	.41489	.4102	.0826	.6095	28
33	.38349	.92354	.41524	.4083	.0828	.6076	27
34	.38376	.92343	.41558	.4063	.0829	.6058	26
35	.38403	.92332	.41592	2.4043	1.0830	2.6040	25
36	.38429	.92321	.41626	.4023	.0832	.6022	24
37	.38456	.92310	.41660	.4004	.0833	.6003	23
38	.38483	.92299	.41694	.3984	.0834	.5985	22
39	.38510	.92287	.41728	.3964	.0836	.5967	21
40	.38537	.92276	.41762	2.3945	1.0837	2.5949	20
41	.38564	.92265	.41797	.3925	.0838	.5931	19
42	.38591	.92254	.41831	.3906	.0840	.5913	18
43	.38617	.92242	.41865	.3886	.0841	.5895	17
44	.38644	.92231	.41899	.3867	.0842	.5877	16
45	.38671	.92220	.41933	2.3847	1.0844	2.5859	15
46	.38698	.92209	.41968	.3828	.0845	.5841	14
47	.38725	.92197	.42002	.3808	.0846	.5823	13
48	.38751	.92186	.42036	.3789	.0847	.5805	12
49	.38778	.92175	.42070	.3770	.0849	.5787	11
50	.38805	.92164	.42105	2.3750	1.0850	2.5770	10
51	.38832	.92152	.42139	.3731	.0851	.5752	9
52	.38859	.92141	.42173	.3712	.0853	.5734	8
53	.38886	.92130	.42207	.3692	.0854	.5716	7
54	.38912	.92118	.42242	.3673	.0855	.5699	6
55	.38939	.92107	.42276	2.3654	1.0857	2.5681	5
56	.38966	.92096	.42310	.3635	.0858	.5663	4
57	.38993	.92084	.42344	.3616	.0859	.5646	3
58	.39019	.92073	.42379	.3597	.0861	.5628	2
59	.39046	.92062	.42413	.3577	.0862	.5610	1
60	.39073	.92050	.42447	2.3558	1.0864	2.5593	0
M	Cosine	Sine	Cotan.	Tan.	Cosec.	Secant	M

67°

23°

M	Sine	Cosine	Tan.	Cotan.	Secant	Cosec.	M
0	.39073	.92050	.42447	2.3558	1.0864	2.5593	60
1	.39100	.92039	.42482	.3539	.0865	.5575	59
2	.39126	.92028	.42516	.3520	.0866	.5558	58
3	.39153	.92016	.42550	.3501	.0868	.5540	57
4	.39180	.92005	.42585	.3482	.0869	.5523	56
5	.39207	.91993	.42619	2.3463	1.0870	2.5506	55
6	.39234	.91982	.42654	.3445	.0872	.5488	54
7	.39260	.91971	.42688	.3426	.0873	.5471	53
8	.39287	.91959	.42722	.3407	.0874	.5453	52
9	.39314	.91948	.42757	.3388	.0876	.5436	51
10	.39341	.91936	.42791	2.3369	1.0877	2.5419	50
11	.39367	.91925	.42826	.3350	.0878	.5402	49
12	.39394	.91913	.42860	.3332	.0880	.5384	48
13	.39421	.91902	.42894	.3313	.0881	.5367	47
14	.39448	.91891	.42929	.3294	.0882	.5350	46
15	.39474	.91879	.42963	2.3276	1.0884	2.5333	45
16	.39501	.91868	.42998	.3257	.0885	.5316	44
17	.39528	.91856	.43032	.3238	.0886	.5299	43
18	.39554	.91845	.43067	.3220	.0888	.5281	42
19	.39581	.91833	.43101	.3201	.0889	.5264	41
20	.39608	.91822	.43136	2.3183	1.0891	2.5247	40
21	.39635	.91810	.43170	.3164	.0892	.5230	39
22	.39661	.91798	.43205	.3145	.0893	.5213	38
23	.39688	.91787	.43239	.3127	.0895	.5196	37
24	.39715	.91775	.43274	.3109	.0896	.5179	36
25	.39741	.91764	.43308	2.3090	1.0897	2.5163	35
26	.39768	.91752	.43343	.3072	.0899	.5146	34
27	.39795	.91741	.43377	.3053	.0900	.5129	33
28	.39821	.91729	.43412	.3035	.0902	.5112	32
29	.39848	.91718	.43447	.3017	.0903	.5095	31
30	.39875	.91706	.43481	2.2998	1.0904	2.5078	30
31	.39901	.91694	.43516	.2980	.0906	.5062	29
32	.39928	.91683	.43550	.2962	.0907	.5045	28
33	.39955	.91671	.43585	.2944	.0908	.5028	27
34	.39981	.91659	.43620	.2925	.0910	.5011	26
35	.40008	.91648	.43654	2.2907	1.0911	2.4995	25
36	.40035	.91636	.43689	.2889	.0913	.4978	24
37	.40061	.91625	.43723	.2871	.0914	.4961	23
38	.40088	.91613	.43758	.2853	.0915	.4945	22
39	.40115	.91601	.43793	.2835	.0917	.4928	21
40	.40141	.91590	.43827	2.2817	1.0918	2.4912	20
41	.40168	.91578	.43862	.2799	.0920	.4895	19
42	.40195	.91566	.43897	.2781	.0921	.4879	18
43	.40221	.91554	.43932	.2763	.0922	.4862	17
44	.40248	.91543	.43966	.2745	.0924	.4846	16
45	.40275	.91531	.44001	2.2727	1.0925	2.4829	15
46	.40301	.91519	.44036	.2709	.0927	.4813	14
47	.40328	.91508	.44070	.2691	.0928	.4797	13
48	.40354	.91496	.44105	.2673	.0929	.4780	12
49	.40381	.91484	.44140	.2655	.0931	.4764	11
50	.40408	.91472	.44175	2.2637	1.0932	2.4748	10
51	.40434	.91461	.44209	.2619	.0934	.4731	9
52	.40461	.91449	.44244	.2602	.0935	.4715	8
53	.40487	.91437	.44279	.2584	.0936	.4699	7
54	.40514	.91425	.44314	.2566	.0938	.4683	6
55	.40541	.91414	.44349	2.2548	1.0939	2.4666	5
56	.40567	.91402	.44383	.2531	.0941	.4650	4
57	.40594	.91390	.44418	.2513	.0942	.4634	3
58	.40620	.91378	.44453	.2495	.0943	.4618	2
59	.40647	.91366	.44488	.2478	.0945	.4602	1
60	.40674	.91354	.44523	2.2460	1.0946	2.4586	0
M	Cosine	Sine	Cotan.	Tan.	Cosec.	Secant	M

66°

24°

M	Sine	Cosine	Tan.	Cotan.	Secant	Cosec.	M
0	.40674	.91354	.44523	2.2460	1.0946	2.4586	60
1	.40700	.91343	.44558	.2443	.0948	.4570	59
2	.40727	.91331	.44593	.2425	.0949	.4554	58
3	.40753	.91319	.44627	.2408	.0951	.4538	57
4	.40780	.91307	.44662	.2390	.0952	.4522	56
5	.40806	.91295	.44697	2.2373	1.0953	2.4506	55
6	.40833	.91283	.44732	.2355	.0955	.4490	54
7	.40860	.91271	.44767	.2338	.0956	.4474	53
8	.40886	.91260	.44802	.2320	.0958	.4458	52
9	.40913	.91248	.44837	.2303	.0959	.4442	51
10	.40939	.91236	.44872	2.2286	1.0961	2.4426	50
11	.40966	.91224	.44907	.2268	.0962	.4411	49
12	.40992	.91212	.44942	.2251	.0963	.4395	48
13	.41019	.91200	.44977	.2234	.0965	.4379	47
14	.41045	.91188	.45012	.2216	.0966	.4363	46
15	.41072	.91176	.45047	2.2199	1.0968	2.4347	45
16	.41098	.91164	.45082	.2182	.0969	.4332	44
17	.41125	.91152	.45117	.2165	.0971	.4316	43
18	.41151	.91140	.45152	.2147	.0972	.4300	42
19	.41178	.91128	.45187	.2130	.0973	.4285	41
20	.41204	.91116	.45222	2.2113	1.0975	2.4269	40
21	.41231	.91104	.45257	.2096	.0976	.4254	39
22	.41257	.91092	.45292	.2079	.0978	.4238	38
23	.41284	.91080	.45327	.2062	.0979	.4222	37
24	.41310	.91068	.45362	.2045	.0981	.4207	36
25	.41337	.91056	.45397	2.2028	1.0982	2.4191	35
26	.41363	.91044	.45432	.2011	.0984	.4176	34
27	.41390	.91032	.45467	.1994	.0985	.4160	33
28	.41416	.91020	.45502	.1977	.0986	.4145	32
29	.41443	.91008	.45537	.1960	.0988	.4130	31
30	.41469	.90996	.45573	2.1943	1.0989	2.4114	30
31	.41496	.90984	.45608	.1926	.0991	.4099	29
32	.41522	.90972	.45643	.1909	.0992	.4083	28
33	.41549	.90960	.45678	.1892	.0994	.4068	27
34	.41575	.90948	.45713	.1875	.0995	.4053	26
35	.41602	.90936	.45748	2.1859	1.0997	2.4037	25
36	.41628	.90924	.45783	.1842	.0998	.4022	24
37	.41654	.90911	.45819	.1825	.1000	.4007	23
38	.41681	.90899	.45854	.1808	.1001	.3992	22
39	.41707	.90887	.45889	.1792	.1003	.3976	21
40	.41734	.90875	.45924	2.1775	1.1004	2.3961	20
41	.41760	.90863	.45960	.1758	.1005	.3946	19
42	.41787	.90851	.45995	.1741	.1007	.3931	18
43	.41813	.90839	.46030	.1725	.1008	.3916	17
44	.41839	.90826	.46065	.1708	.1010	.3901	16
45	.41866	.90814	.46101	2.1692	1.1011	2.3886	15
46	.41892	.90802	.46136	.1675	.1013	.3871	14
47	.41919	.90790	.46171	.1658	.1014	.3856	13
48	.41945	.90778	.46206	.1642	.1016	.3841	12
49	.41972	.90765	.46242	.1625	.1017	.3826	11
50	.41998	.90753	.46277	2.1609	1.1019	2.3811	10
51	.42024	.90741	.46312	.1592	.1020	.3796	9
52	.42051	.90729	.46348	.1576	.1022	.3781	8
53	.42077	.90717	.46383	.1559	.1023	.3766	7
54	.42103	.90704	.46418	.1543	.1025	.3751	6
55	.42130	.90692	.46454	2.1527	1.1026	2.3736	5
56	.42156	.90680	.46489	.1510	.1028	.3721	4
57	.42183	.90668	.46524	.1494	.1029	.3706	3
58	.42209	.90655	.46560	.1478	.1031	.3691	2
59	.42235	.90643	.46595	.1461	.1032	.3677	1
60	.42262	.90631	.46631	2.1445	1.1034	2.3662	0
M	Cosine	Sine	Cotan.	Tan.	Cosec.	Secant	M

65°

25°

M	Sine	Cosine	Tan.	Cotan.	Secant	Cosec.	M
0	.42262	.90631	.46631	2.1445	1.1034	2.3662	60
1	.42288	.90618	.46666	.1429	.1035	.3647	59
2	.42314	.90606	.46702	.1412	.1037	.3632	58
3	.42341	.90594	.46737	.1396	.1038	.3618	57
4	.42367	.90581	.46772	.1380	.1040	.3603	56
5	.42394	.90569	.46808	2.1364	1.1041	2.3588	55
6	.42420	.90557	.46843	.1348	.1043	.3574	54
7	.42446	.90544	.46879	.1331	.1044	.3559	53
8	.42473	.90532	.46914	.1315	.1046	.3544	52
9	.42499	.90520	.46950	.1299	.1047	.3530	51
10	.42525	.90507	.46985	2.1283	1.1049	2.3515	50
11	.42552	.90495	.47021	.1267	.1050	.3501	49
12	.42578	.90483	.47056	.1251	.1052	.3486	48
13	.42604	.90470	.47092	.1235	.1053	.3472	47
14	.42630	.90458	.47127	.1219	.1055	.3457	46
15	.42657	.90445	.47163	2.1203	1.1056	2.3443	45
16	.42683	.90433	.47199	.1187	.1058	.3428	44
17	.42709	.90421	.47234	.1171	.1059	.3414	43
18	.42736	.90408	.47270	.1155	.1061	.3399	42
19	.42762	.90396	.47305	.1139	.1062	.3385	41
20	.42788	.90383	.47341	2.1123	1.1064	2.3371	40
21	.42815	.90371	.47376	.1107	.1065	.3356	39
22	.42841	.90358	.47412	.1092	.1067	.3342	38
23	.42867	.90346	.47448	.1076	.1068	.3328	37
24	.42893	.90333	.47483	.1060	.1070	.3313	36
25	.42920	.90321	.47519	2.1044	1.1072	2.3299	35
26	.42946	.90308	.47555	.1028	.1073	.3285	34
27	.42972	.90296	.47590	.1013	.1075	.3271	33
28	.42998	.90283	.47626	.0997	.1076	.3256	32
29	.43025	.90271	.47662	.0981	.1078	.3242	31
30	.43051	.90258	.47697	2.0965	1.1079	2.3228	30
31	.43077	.90246	.47733	.0950	.1081	.3214	29
32	.43104	.90233	.47769	.0934	.1082	.3200	28
33	.43130	.90221	.47805	.0918	.1084	.3186	27
34	.43156	.90208	.47840	.0903	.1085	.3172	26
35	.43182	.90196	.47876	2.0887	1.1087	2.3158	25
36	.43208	.90183	.47912	.0872	.1088	.3143	24
37	.43235	.90171	.47948	.0856	.1090	.3129	23
38	.43261	.90158	.47983	.0840	.1092	.3115	22
39	.43287	.90145	.48019	.0825	.1093	.3101	21
40	.43313	.90133	.48055	2.0809	1.1095	2.3087	20
41	.43340	.90120	.48091	.0794	1096	.3073	19
42	.43366	.90108	.48127	.0778	.1098	.3059	18
43	.43392	.90095	.48162	.0763	.1099	.3046	17
44	.43418	.90082	.48198	.0747	.1101	.3032	16
45	.43444	.90070	.48234	2.0732	1.1102	2.3018	15
46	.43471	.90057	.48270	.0717	.1104	.3004	14
47	.43497	.90044	.48306	.0701	.1106	.2990	13
48	.43523	.90032	.48342	.0686	.1107	.2976	12
49	.43549	.90019	.48378	.0671	.1109	.2962	11
50	.43575	.90006	.48414	2.0655	1.1110	2.2949	10
51	.43602	.89994	.48449	.0640	.1112	.2935	9
52	.43628	.89981	.48485	.0625	.1113	.2921	8
53	.43654	.89968	.48521	.0609	.1115	.2907	7
54	.43680	.89956	.48557	.0594	.1116	.2894	6
55	.43706	.89943	.48593	2.0579	1.1118	2.2880	5
56	.43732	.89930	.48629	.0564	.1120	.2866	4
57	.43759	.89918	.48665	.0548	.1121	.2853	3
58	.43785	.89905	.48701	.0533	.1123	.2839	2
59	.43811	.89892	.48737	.0518	.1124	.2825	1
60	.43837	.89879	.48773	2.0503	1.1126	2.2812	0
M	Cosine	Sine	Cotan.	Tan.	Cosec.	Secant	M

64°

26°

M	Sine	Cosine	Tan.	Cotan.	Secant	Cosec.	M
0	.43837	.89879	.48773	2.0503	1.1126	2.2812	60
1	.43863	.89867	.48809	.0488	.1127	.2798	59
2	.43889	.89854	.48845	.0473	.1129	.2784	58
3	.43915	.89841	.48881	.0458	.1131	.2771	57
4	.43942	.89828	.48917	.0443	.1132	.2757	56
5	.43968	.89815	.48953	2.0427	1.1134	2.2744	55
6	.43994	.89803	.48989	.0412	.1135	.2730	54
7	.44020	.89790	.49025	.0397	.1137	.2717	53
8	.44046	.89777	.49062	2.0382	.1139	.2703	52
9	.44072	.89764	.49098	.0367	.1140	.2690	51
10	.44098	.89751	.49134	2.0352	1.1142	2.2676	50
11	.44124	.89739	.49170	.0338	.1143	.2663	49
12	.44150	.89726	.49206	.0323	.1145	.2650	48
13	.44177	.89713	.49242	.0308	.1147	.2636	47
14	.44203	.89700	.49278	.0293	.1148	.2623	46
15	.44229	.89687	.49314	2.0278	1.1150	2.2610	45
16	.44255	.89674	.49351	.0263	.1151	.2596	44
17	.44281	.89661	.49387	.0248	.1153	.2583	43
18	.44307	.89649	.49423	.0233	.1155	.2570	42
19	.44333	.89636	.49459	.0219	.1156	.2556	41
20	.44359	.89623	.49495	2.0204	1.1158	2.2543	40
21	.44385	.89610	.49532	.0189	.1159	.2530	39
22	.44411	.89597	.49568	.0174	.1161	.2517	38
23	.44437	.89584	.49604	.0159	.1163	.2503	37
24	.44463	.89571	.49640	.0145	.1164	.2490	36
25	.44489	.89558	.49677	2.0130	1.1166	2.2477	35
26	.44516	.89545	.49713	.0115	.1167	.2464	34
27	.44542	.89532	.49749	.0101	.1169	.2451	33
28	.44568	.89519	.49785	.0086	.1171	.2438	32
29	.44594	.89506	.49822	.0071	.1172	.2425	31
30	.44620	.89493	.49858	2.0057	1.1174	2.2411	30
31	.44646	.89480	.49894	.0042	.1176	.2398	29
32	.44672	.89467	.49931	.0028	.1177	.2385	28
33	.44698	.89454	.49967	.0013	.1179	.2372	27
34	.44724	.89441	.50003	1.9998	.1180	.2359	26
35	.44750	.89428	.50040	1.9984	1.1182	2.2348	25
36	.44776	.89415	.50076	.9969	.1184	.2333	24
37	.44802	.89402	.50113	.9955	.1185	.2320	23
38	.44828	.89389	.50149	.9940	.1187	.2307	22
39	.44854	.89376	.50185	.9926	.1189	.2294	21
40	.44880	.89363	.50222	1.9912	1.1190	2.2282	20
41	.44906	.89350	.50258	.9897	.1192	.2269	19
42	.44932	.89337	.50295	.9883	.1193	.2256	18
43	.44958	.89324	.50331	.9868	.1195	.2243	17
44	.44984	.89311	.50368	.9854	.1197	.2230	16
45	.45010	.89298	.50404	1.9840	1.1198	2.2217	15
46	.45036	.89285	.50441	.9825	.1200	.2204	14
47	.45062	.89272	.50477	.9811	.1202	.2192	13
48	.45088	.89258	.50514	.9797	.1203	.2179	12
49	.45114	.89245	.50550	.9782	.1205	.2166	11
50	.45140	.89232	.50587	1.9768	1.1207	2.2153	10
51	.45166	.89219	.50623	.9754	.1208	.2141	9
52	.45191	.89206	.50660	.9739	.1210	.2128	8
53	.45217	.89193	.50696	.9725	.1212	.2115	7
54	.45243	.89180	.50733	.9711	.1213	.2103	6
55	.45269	.89166	.50769	1.9697	1.1215	2.2090	5
56	.45295	.89153	.50806	.9683	.1217	.2077	4
57	.45321	.89140	.50843	.9668	.1218	.2065	3
58	.45347	.89127	.50879	.9654	.1220	.2052	2
59	.45373	.89114	.50916	.9640	.1222	.2039	1
60	.45399	.89101	.50952	1.9626	1.1223	2.2027	0
M	Cosine	Sine	Cotan.	Tan.	Cosec.	Secant	M

63°

27°

M	Sine	Cosine	Tan.	Cotan.	Secant	Cosec.	M
0	.45399	.89101	.50952	1.9626	1.1223	2.2027	60
1	.45425	.89087	.50989	.9612	.1225	.2014	59
2	.45451	.89074	.51026	.9598	.1226	.2002	58
3	.45477	.89061	.51062	.9584	.1228	.1989	57
4	.45503	.89048	.51099	.9570	.1230	.1977	56
5	.45528	.89034	.51136	1.9556	1.1231	2.1964	55
6	.45554	.89021	.51172	.9542	.1233	.1952	54
7	.45580	.89008	.51209	.9528	.1235	.1939	53
8	.45606	.88995	.51246	.9514	.1237	.1927	52
9	.45632	.88981	.51283	.9500	.1238	.1914	51
10	.45658	.88968	.51319	1.9486	1.1240	2.1902	50
11	.45684	.88955	.51356	.9472	.1242	.1889	49
12	.45710	.88942	.51393	.9458	.1243	.1877	48
13	.45736	.88928	.51430	.9444	.1245	.1865	47
14	.45761	.88915	.51466	.9430	.1247	.1852	46
15	.45787	.88902	.51503	1.9416	1.1248	2.1840	45
16	.45813	.88888	.51540	.9402	.1250	.1828	44
17	.45839	.88875	.51577	.9388	.1252	.1815	43
18	.45865	.88862	.51614	.9375	.1253	.1803	42
19	.45891	.88848	.51651	.9361	.1255	.1791	41
20	.45917	.88835	.51687	1.9347	1.1257	2.1778	40
21	.45942	.88822	.51724	.9333	.1258	.1766	39
22	.45968	.88808	.51761	.9319	.1260	.1754	38
23	.45994	.88795	.51798	.9306	.1262	.1742	37
24	.46020	.88781	.51835	.9292	.1264	.1730	36
25	.46046	.88768	.51872	1.9278	1.1265	2.1717	35
26	.46072	.88755	.51909	.9264	.1267	.1705	34
27	.46097	.88741	.51946	.9251	.1269	.1693	33
28	.46123	.88728	.51983	.9237	.1270	.1681	32
29	.46149	.88714	.52020	.9223	.1272	.1669	31
30	.46175	.88701	.52057	1.9210	1.1274	2.1657	30
31	.46201	.88688	.52094	.9196	.1275	.1645	29
32	.46226	.88674	.52131	.9182	.1277	.1633	28
33	.46252	.88661	.52168	.9169	.1279	.1620	27
34	.46278	.88647	.52205	.9155	.1281	.1608	26
35	.46304	.88634	.52242	1.9142	1.1282	2.1596	25
36	.46330	.88620	.52279	.9128	.1284	.1584	24
37	.46355	.88607	.52316	.9115	.1286	.1572	23
38	.46381	.88593	.52353	.9101	.1287	.1560	22
39	.46407	.88580	.52390	.9088	.1289	.1548	21
40	.46433	.88566	.52427	1.9074	1.1291	2.1536	20
41	.46458	.88553	.52464	.9061	.1293	.1525	19
42	.46484	.88539	.52501	.9047	.1294	.1513	18
43	.46510	.88526	.52538	.9034	.1296	.1501	17
44	.46536	.88512	.52575	.9020	.1298	.1489	16
45	.46561	.88499	.52612	1.9007	1.1299	2.1477	15
46	.46587	.88485	.52650	.8993	.1301	.1465	14
47	.46613	.88472	.52687	.8980	.1303	.1453	13
48	.46639	.88458	.52724	.8967	.1305	.1441	12
49	.46664	.88444	.52761	.8953	.1306	.1430	11
50	.46690	.88431	.52798	1.8940	1.1308	2.1418	10
51	.46716	.88417	.52836	.8927	.1310	.1406	9
52	.46741	.88404	.52873	.8913	.1312	.1394	8
53	.46767	.88390	.52910	.8900	.1313	.1382	7
54	.46793	.88376	.52947	.8887	.1315	.1371	6
55	.46819	.88363	.52984	1.8873	1.1317	2.1359	5
56	.46844	.88349	.53022	.8860	.1319	.1347	4
57	.46870	.88336	.53059	.8847	.1320	.1335	3
58	.46896	.88322	.53096	.8834	.1322	.1324	2
59	.46921	.88308	.53134	.8820	.1324	.1312	1
60	.46947	.88295	.53171	1.8807	1.1326	2.1300	0
M	Cosine	Sine	Cotan.	Tan.	Cosec.	Secant	M

62°

28°

M	Sine	Cosine	Tan.	Cotan.	Secant	Cosec.	M
0	.46947	.88295	.53171	1.8807	1.1326	2.1300	60
1	.46973	.88281	.53208	.8794	.1327	.1289	59
2	.46998	.88267	.53245	.8781	.1329	.1277	58
3	.47024	.88254	.53283	.8768	.1331	.1266	57
4	.47050	.88240	.53320	.8754	.1333	.1254	56
5	.47075	.88226	.53358	1.8741	1.1334	2.1242	55
6	.47101	.88213	.53395	.8728	.1336	.1231	54
7	.47127	.88199	.53432	.8715	.1338	.1219	53
8	.47152	.88185	.53470	.8702	.1340	.1208	52
9	.47178	.88171	.53507	.8689	.1341	.1196	51
10	.47204	.88158	.53545	1.8676	1.1343	2.1185	50
11	.47229	.88144	.53582	.8663	.1345	.1173	49
12	.47255	.88130	.53619	.8650	.1347	.1162	48
13	.47281	.88117	.53657	.8637	.1349	.1150	47
14	.47306	.88103	.53694	.8624	.1350	.1139	46
15	.47332	.88089	.53732	1.8611	1.1352	2.1127	45
16	.47357	.88075	.53769	.8598	.1354	.1116	44
17	.47383	.88061	.53807	.8585	.1356	.1104	43
18	.47409	.88048	.53844	.8572	.1357	.1093	42
19	.47434	.88034	.53882	.8559	.1359	.1082	41
20	.47460	.88020	.53919	1.8546	1.1361	2.1070	40
21	.47486	.88006	.53957	.8533	.1363	.1059	39
22	.47511	.87992	.53995	.8520	.1365	.1048	38
23	.47537	.87979	.54032	.8507	.1366	.1036	37
24	.47562	.87965	.54070	.8495	.1368	.1025	36
25	.47588	.87951	.54107	1.8482	1.1370	2.1014	35
26	.47613	.87937	.54145	.8469	.1372	.1002	34
27	.47639	.87923	.54183	.8456	.1373	.0991	33
28	.47665	.87909	.54220	.8443	.1375	.0980	32
29	.47690	.87895	.54258	.8430	.1377	.0969	31
30	.47716	.87882	.54295	1.8418	1.1379	2.0957	30
31	.47741	.87868	.54333	.8405	.1381	.0946	29
32	.47767	.87854	.54371	.8392	.1382	.0935	28
33	.47792	.87840	.54409	.8379	.1384	.0924	27
34	.47818	.87826	.54446	.8367	.1386	.0912	26
35	.47844	.87812	.54484	1.8354	1.1388	2.0901	25
36	.47869	.87798	.54522	.8341	.1390	.0890	24
37	.47895	.87784	.54559	.8329	.1391	.0879	23
38	.47920	.87770	.54597	.8316	.1393	.0868	22
39	.47946	.87756	.54635	.8303	.1395	.0857	21
40	.47971	.87742	.54673	1.8291	1.1397	2.0846	20
41	.47997	.87728	.54711	.8278	.1399	.0835	19
42	.48022	.87715	.54748	.8265	.1401	.0824	18
43	.48048	.87701	.54786	.8253	.1402	.0812	17
44	.48073	.87687	.54824	.8240	.1404	.0801	16
45	.48099	.87673	.54862	1.8227	1.1406	2.0790	15
46	.48124	.87659	.54900	.8215	.1408	.0779	14
47	.48150	.87645	.54937	.8202	.1410	.0768	13
48	.48175	.87631	.54975	.8190	.1411	.0757	12
49	.48201	.87617	.55013	.8177	.1413	.0746	11
50	.48226	.87603	.55051	1.8165	1.1415	2.0735	10
51	.48252	.87588	.55089	.8152	.1417	.0725	9
52	.48277	.87574	.55127	.8140	.1419	.0714	8
53	.48303	.87560	.55165	.8127	.1421	.0703	7
54	.48328	.87546	.55203	.8115	.1422	.0692	6
55	.48354	.87532	.55241	1.8102	1.1424	2.0681	5
56	.48379	.87518	.55279	.8090	.1426	.0670	4
57	.48405	.87504	.55317	.8078	.1428	.0659	3
58	.48430	.87490	.55355	.8065	.1430	.0648	2
59	.48455	.87476	.55393	.8053	.1432	.0637	1
60	.48481	.87462	.55431	1.8040	1.1433	2.0627	0
M	Cosine	Sine	Cotan.	Tan.	Cosec.	Secant	M

61°

29°

M	Sine	Cosine	Tan.	Cotan.	Secant	Cosec.	M
0	.48481	.87462	.55431	1.8040	1.1433	2.0627	60
1	.48506	.87448	.55469	.8028	.1435	.0616	59
2	.48532	.87434	.55507	.8016	.1437	.0605	58
3	.48557	.87420	.55545	.8003	.1439	.0594	57
4	.48583	.87405	.55583	.7991	.1441	.0583	56
5	.48608	.87391	.55621	1.7979	1.1443	2.0573	55
6	.48633	.87377	.55659	.7966	.1445	.0562	54
7	.48659	.87363	.55697	.7954	.1446	.0551	53
8	.48684	.87349	.55735	.7942	.1448	.0540	52
9	.48710	.87335	.55774	.7930	.1450	.0530	51
10	.48735	.87320	.55812	1.7917	1.1452	2.0519	50
11	.48760	.87306	.55850	.7905	.1454	.0508	49
12	.48786	.87292	.55888	.7893	.1456	.0498	48
13	.48811	.87278	.55926	.7881	.1458	.0487	47
14	.48837	.87264	.55964	.7868	.1459	.0476	46
15	.48862	.87250	.56003	1.7856	1.1461	2.0466	45
16	.48887	.87235	.56041	.7844	.1463	.0455	44
17	.48913	.87221	.56079	.7832	.1465	.0444	43
18	.48938	.87207	.56117	.7820	.1467	.0434	42
19	.48964	.87193	.56156	.7808	.1469	.0423	41
20	.48989	.87178	.56194	1.7795	1.1471	2.0413	40
21	.49014	.87164	.56232	.7783	.1473	.0402	39
22	.49040	.87150	.56270	.7771	.1474	.0392	38
23	.49065	.87136	.56309	.7759	.1476	.0381	37
24	.49090	.87121	.56347	.7747	.1478	.0370	36
25	.49116	.87107	.56385	1.7735	1.1480	2.0360	35
26	.49141	.87093	.56424	.7723	.1482	.0349	34
27	.49166	.87078	.56462	.7711	.1484	.0339	33
28	.49192	.87064	.56500	.7699	.1486	.0329	32
29	.49217	.87050	.56539	.7687	.1488	.0318	31
30	.49242	.87035	.56577	1.7675	1.1489	2.0308	30
31	.49268	.87021	.56616	.7663	.1491	.0297	29
32	.49293	.87007	.56654	.7651	.1493	.0287	28
33	.49318	.86992	.56692	.7639	.1495	.0276	27
34	.49343	.86978	.56731	.7627	.1497	.0266	26
35	.49369	.86964	.56769	1.7615	1.1499	2.0256	25
36	.49394	.86949	.56808	.7603	.1501	.0245	24
37	.49419	.86935	.56846	.7591	.1503	.0235	23
38	.49445	.86921	.56885	.7579	.1505	.0224	22
39	.49470	.86906	.56923	.7567	.1507	.0214	21
40	.49495	.86892	.56962	1.7555	1.1508	2.0204	20
41	.49521	.86877	.57000	.7544	.1510	.0194	19
42	.49546	.86863	.57039	.7532	.1512	.0183	18
43	.49571	.86849	.57077	.7520	.1514	.0173	17
44	.49596	.86834	.57116	.7508	.1516	.0163	16
45	.49622	.86820	.57155	1.7496	1.1518	2.0152	15
46	.49647	.86805	.57193	.7484	.1520	.0142	14
47	.49672	.86791	.57232	.7473	.1522	.0132	13
48	.49697	.86776	.57270	.7461	.1524	.0122	12
49	.49723	.86762	.57309	.7449	.1526	.0111	11
50	.49748	.86748	.57348	1.7437	1.1528	2.0101	10
51	.49773	.86733	.57386	.7426	.1530	.0091	9
52	.49798	.86719	.57425	.7414	.1531	.0081	8
53	.49823	.86704	.57464	.7402	.1533	.0071	7
54	.49849	.86690	.57502	.7390	.1535	.0061	6
55	.49874	.86675	.57541	1.7379	1.1537	2.0050	5
56	.49899	.86661	.57580	.7367	.1539	.0040	4
57	.49924	.86646	.57619	.7355	.1541	.0030	3
58	.49950	.86632	.57657	.7344	.1543	.0020	2
59	.49975	.86617	.57696	.7332	.1545	.0010	1
60	.50000	.86603	.57735	1.7320	1.1547	2.0000	0
M	Cosine	Sine	Cotan.	Tan.	Cosec.	Secant	M

60°

30°

M	Sine	Cosine	Tan.	Cotan.	Secant	Cosec.	M
0	.50000	.86603	.57735	1.7320	1.1547	2.0000	60
1	.50025	.86588	.57774	.7309	.1549	1.9990	59
2	.50050	.86573	.57813	.7297	.1551	.9980	58
3	.50075	.86559	.57851	.7286	.1553	.9970	57
4	.50101	.86544	.57890	.7274	.1555	.9960	56
5	.50126	.86530	.57929	1.7262	1.1557	1.9950	55
6	.50151	.86515	.57968	.7251	.1559	.9940	54
7	.50176	.86500	.58007	.7239	.1561	.9930	53
8	.50201	.86486	.58046	.7228	.1562	.9920	52
9	.50226	.86471	.58085	.7216	.1564	.9910	51
10	.50252	.86457	.58123	1.7205	1.1566	1.9900	50
11	.50277	.86442	.58162	.7193	.1568	.9890	49
12	.50302	.86427	.58201	.7182	.1570	.9880	48
13	.50327	.86413	.58240	.7170	.1572	.9870	47
14	.50352	.86398	.58279	.7159	.1574	.9860	46
15	.50377	.86383	.58318	1.7147	1.1576	1.9850	45
16	.50402	.86369	.58357	.7136	.1578	.9840	44
17	.50428	.86354	.58396	.7124	.1580	.9830	43
18	.50453	.86339	.58435	.7113	.1582	.9820	42
19	.50478	.86325	.58474	.7101	.1584	.9811	41
20	.50503	.86310	.58513	1.7090	1.1586	1.9801	40
21	.50528	.86295	.58552	.7079	.1588	.9791	39
22	.50553	.86281	.58591	.7067	.1590	.9781	38
23	.50578	.86266	.58630	.7056	.1592	.9771	37
24	.50603	.86251	.58670	.7044	.1594	.9761	36
25	.50628	.86237	.58709	1.7033	1.1596	1.9752	35
26	.50653	.86222	.58748	.7022	.1598	.9742	34
27	.50679	.86207	.58787	.7010	.1600	.9732	33
28	.50704	.86192	.58826	.6999	.1602	.9722	32
29	.50729	.86178	.58865	.6988	.1604	.9713	31
30	.50754	.86163	.58904	1.6977	1.1606	1.9703	30
31	.50779	.86148	.58944	.6965	.1608	.9693	29
32	.50804	.86133	.58983	.6954	.1610	.9683	28
33	.50829	.86118	.59022	.6943	.1612	.9674	27
34	.50854	.86104	.59061	.6931	.1614	.9664	26
35	.50879	.86089	.59100	1.6920	1.1616	1.9654	25
36	.50904	.86074	.59140	.6909	.1618	.9645	24
37	.50929	.86059	.59179	.6898	.1620	.9635	23
38	.50954	.86044	.59218	.6887	.1622	.9625	22
39	.50979	.86030	.59258	.6875	.1624	.9616	21
40	.51004	.86015	.59297	1.6864	1.1626	1.9606	20
41	.51029	.86000	.59336	.6853	.1628	.9596	19
42	.51054	.85985	.59376	.6842	.1630	9587	18
43	.51079	.85970	.59415	.6831	.1632	9577	17
44	.51104	.85955	.59454	.6820	.1634	.9568	16
45	.51129	.85941	.59494	1.6808	1.1636	1.9558	15
46	.51154	.85926	.59533	.6797	.1638	.9549	14
47	.51179	.85911	.59572	.6786	.1640	.9539	13
48	.51204	.85896	.59612	.6775	.1642	.9530	12
49	.51229	.85881	.59651	.6764	.1644	.9520	11
50	.51254	.85866	.59691	1.6753	1.1646	1.9510	10
51	.51279	.85851	.59730	.6742	.1648	.9501	9
52	.51304	.85836	.59770	.6731	.1650	.9491	8
53	.51329	.85821	.59809	.6720	.1652	.9482	7
54	.51354	.85806	.59849	.6709	.1654	.9473	6
55	.51379	.85791	.59888	1.6698	1.1656	1.9463	5
56	.51404	.85777	.59928	.6687	.1658	.9454	4
57	.51429	.85762	.59967	.6676	.1660	.9444	3
58	.51454	.85747	.60007	.6665	.1662	.9435	2
59	.51479	.85732	.60046	.6654	.1664	.9425	1
60	.51504	.85717	.60086	1.6643	1.1666	1.9416	0
M	Cosine	Sine	Cotan.	Tan.	Cosec.	Secant	M

59°

31°

M	Sine	Cosine	Tan.	Cotan.	Secant	Cosec.	M
0	.51504	.85717	.60086	1.6643	1.1666	1.9416	60
1	.51529	.85702	.60126	.6632	.1668	.9407	59
2	.51554	.85687	.60165	.6621	.1670	.9397	58
3	.51578	.85672	.60205	.6610	.1672	.9388	57
4	.51603	.85657	.60244	.6599	.1674	.9378	56
5	.51628	.85642	.60284	1.6588	1.1676	1.9369	55
6	.51653	.85627	.60324	.6577	.1678	.9360	54
7	.51678	.85612	.60363	.6566	.1681	.9350	53
8	.51703	.85597	.60403	.6555	.1683	.9341	52
9	.51728	.85582	.60443	.6544	.1685	.9332	51
10	.51753	.85566	.60483	1.6534	1.1687	1.9322	50
11	.51778	.85551	.60522	.6523	.1689	.9313	49
12	.51803	.85536	.60562	.6512	.1691	.9304	48
13	.51827	.85521	.60602	.6501	.1693	.9295	47
14	.51852	.85506	.60642	.6490	.1695	.9285	46
15	.51877	.85491	.60681	1.6479	1.1697	1.9276	45
16	.51902	.85476	.60721	.6469	.1699	.9267	44
17	.51927	.85461	.60761	.6458	.1701	.9258	43
18	.51952	.85446	.60801	.6447	.1703	.9248	42
19	.51977	.85431	.60841	.6436	.1705	.9239	41
20	.52002	.85416	.60881	1.6425	1.1707	1.9230	40
21	.52026	.85400	.60920	.6415	.1709	.9221	39
22	.52051	.85385	.60960	.6404	.1712	.9212	38
23	.52076	.85370	.61000	.6393	.1714	.9203	37
24	.52101	.85355	.61040	.6383	.1716	.9193	36
25	.52126	.85340	.61080	1.6372	1.1718	1.9184	35
26	.52151	.85325	.61120	.6361	.1720	.9175	34
27	.52175	.85309	.61160	.6350	.1722	.9166	33
28	.52200	.85294	.61200	.6340	.1724	.9157	32
29	.52225	.85279	.61240	.6329	.1726	.9148	31
30	.52250	.85264	.61280	1.6318	1.1728	1.9139	30
31	.52275	.85249	.61320	.6308	.1730	.9130	29
32	.52299	.85234	.61360	.6297	.1732	.9121	28
33	.52324	.85218	.61400	.6286	.1734	.9112	27
34	.52349	.85203	.61440	.6276	.1737	.9102	26
35	.52374	.85188	.61480	1.6265	1.1739	1.9093	25
36	.52398	.85173	.61520	.6255	.1741	.9084	24
37	.52423	.85157	.61560	.6244	.1743	.9075	23
38	.52448	.85142	.61601	.6233	.1745	.9066	22
39	.52473	.85127	.61641	.6223	.1747	.9057	21
40	.52498	.85112	.61681	1.6212	1.1749	1.9048	20
41	.52522	.85096	.61721	.6202	.1751	.9039	19
42	.52547	.85081	.61761	.6191	.1753	.9030	18
43	.52572	.85066	.61801	.6181	.1756	.9021	17
44	.52597	.85050	.61842	.6170	.1758	.9013	16
45	.52621	.85035	.61882	1.6160	1.1760	1.9004	15
46	.52646	.85020	.61922	.6149	.1762	.8995	14
47	.52671	.85004	.61962	.6139	.1764	.8986	13
48	.52695	.84989	.62003	.6128	.1766	.8977	12
49	.52720	.84974	.62043	.6118	.1768	.8968	11
50	.52745	.84959	.62083	1.6107	1.1770	1.8959	10
51	.52770	.84943	.62123	.6097	.1772	.8950	9
52	.52794	.84928	.62164	.6086	.1775	.8941	8
53	.52819	.84912	.62204	.6076	.1777	.8932	7
54	.52844	.84897	.62244	.6066	.1779	.8924	6
55	.52868	.84882	.62285	1.6055	1.1781	1.8915	5
56	.52893	.84866	.62325	.6045	.1783	.8906	4
57	.52918	.84851	.62366	.6034	.1785	.8897	3
58	.52942	.84836	.62406	.6024	.1787	.8888	2
59	.52967	.84820	.62446	.6014	.1790	.8879	1
60	.52992	.84805	.62487	1.6003	1.1792	1.8871	0
M	Cosine	Sine	Cotan.	Tan.	Cosec.	Secant	M

58°

32°

M	Sine	Cosine	Tan.	Cotan.	Secant	Cosec.	M
0	.52992	.84805	.62487	1.6003	1.1792	1.8871	60
1	.53016	.84789	.62527	.5993	.1794	.8862	59
2	.53041	.84774	.62568	.5983	.1796	.8853	58
3	.53066	.84758	.62608	.5972	.1798	.8844	57
4	.53090	.84743	.62649	.5962	.1800	.8836	56
5	.53115	.84728	.62689	1.5952	1.1802	1.8827	55
6	.53140	.84712	.62730	.5941	.1805	.8818	54
7	.53164	.84697	.62770	.5931	.1807	.8809	53
8	.53189	.84681	.62811	.5921	.1809	.8801	52
9	.53214	.84666	.62851	.5910	.1811	.8792	51
10	.53238	.84650	.62892	1.5900	1.1813	1.8783	50
11	.53263	.84635	.62933	.5890	.1815	.8775	49
12	.53288	.84619	.62973	.5880	.1818	.8766	48
13	.53312	.84604	.63014	.5869	.1820	.8757	47
14	.53337	.84588	.63055	.5859	.1822	.8749	46
15	.53361	.84573	.63095	1.5849	1.1824	1.8740	45
16	.53386	.84557	.63136	.5839	.1826	.8731	44
17	.53411	.84542	.63177	.5829	.1828	.8723	43
18	.53435	.84526	.63217	.5818	.1831	.8714	42
19	.53460	.84511	.63258	.5808	.1833	.8706	41
20	.53484	.84495	.63299	1.5798	1.1835	1.8697	40
21	.53509	.84479	.63339	.5788	.1837	.8688	39
22	.53533	.84464	.63380	.5778	.1839	.8680	38
23	.53558	.84448	.63421	.5768	.1841	.8671	37
24	.53583	.84433	.63462	.5757	.1844	.8663	36
25	.53607	.84417	.63503	1.5747	1.1846	1.8654	35
26	.53632	.84402	.63543	.5737	.1848	.8646	34
27	.53656	.84386	.63584	.5727	.1850	.8637	33
28	.53681	.84370	.63625	.5717	.1852	.8629	32
29	.53705	.84355	.63666	.5707	.1855	.8620	31
30	.53730	.84339	.63707	1.5697	1.1857	1.8611	30
31	.53754	.84323	.63748	.5687	.1859	.8603	29
32	.53779	.84308	.63789	.5677	.1861	.8595	28
33	.53803	.84292	.63830	.5667	.1863	.8586	27
34	.53828	.84276	.63871	.5657	.1866	.8578	26
35	.53852	.84261	.63912	1.5646	1.1868	1.8569	25
36	.53877	.84245	.63953	.5636	.1870	.8561	24
37	.53901	.84229	.63994	.5626	.1872	.8552	23
38	.53926	.84214	.64035	.5616	.1874	.8544	22
39	.53950	.84198	.64076	.5606	.1877	.8535	21
40	.53975	.84182	.64117	1.5596	1.1879	1.8527	20
41	.53999	.84167	.64158	.5586	.1881	.8519	19
42	.54024	.84151	.64199	.5577	.1883	.8510	18
43	.54048	.84135	.64240	.5567	.1886	.8502	17
44	.54073	.84120	.64281	.5557	.1888	.8493	16
45	.54097	.84104	.64322	1.5547	1.1890	1.8485	15
46	.54122	.84088	.64363	.5537	.1892	.8477	14
47	.54146	.84072	.64404	.5527	.1894	.8468	13
48	.54171	.84057	.64446	.5517	.1897	.8460	12
49	.54195	.84041	.64487	.5507	.1899	.8452	11
50	.54220	.84025	.64528	1.5497	1.1901	1.8443	10
51	.54244	.84009	.64569	.5487	.1903	.8435	9
52	.54268	.83993	.64610	.5477	.1906	.8427	8
53	.54293	.83978	.64652	.5467	.1908	.8418	7
54	.54317	.83962	.64693	.5458	.1910	.8410	6
55	.54342	.83946	.64734	1.5448	1.1912	1.8402	5
56	.54366	.83930	.64775	.5438	.1915	.8394	4
57	.54391	.83914	.64817	.5428	.1917	.8385	3
58	.54415	.83899	.64858	.5418	.1919	.8377	2
59	.54439	.83883	.64899	.5408	.1921	.8369	1
60	.54464	.83867	.64941	1.5399	1.1924	1.8361	0
M	Cosine	Sine	Cotan.	Tan.	Cosec.	Secant	M

57°

33°

M	Sine	Cosine	Tan.	Cotan.	Secant	Cosec.	M
0	.54464	.83867	.64941	1.5399	1.1924	1.8361	60
1	.54488	.83851	.64982	.5389	.1926	.8352	59
2	.54513	.83835	.65023	.5379	.1928	.8344	58
3	.54537	.83819	.65065	.5369	.1930	.8336	57
4	.54561	.83804	.65106	.5359	.1933	.8328	56
5	.54586	.83788	.65148	1.5350	1.1935	1.8320	55
6	.54610	.83772	.65189	.5340	.1937	.8311	54
7	.54634	.83756	.65231	.5330	.1939	.8303	53
8	.54659	.83740	.65272	.5320	.1942	.8295	52
9	.54683	.83724	.65314	.5311	.1944	.8287	51
10	.54708	.83708	.65355	1.5301	1.1946	1.8279	50
11	.54732	.83692	.65397	.5291	.1948	.8271	49
12	.54756	.83676	.65438	.5282	.1951	.8263	48
13	.54781	.83660	.65480	.5272	.1953	.8255	47
14	.54805	.83644	.65521	.5262	.1955	.8246	46
15	.54829	.83629	.65563	1.5252	1.1958	1.8238	45
16	.54854	.83613	.65604	.5243	.1960	.8230	44
17	.54878	.83597	.65646	.5233	.1962	.8222	43
18	.54902	.83581	.65688	.5223	.1964	.8214	42
19	.54926	.83565	.65729	.5214	.1967	.8206	41
20	.54951	.83549	.65771	1.5204	1.1969	1.8198	40
21	.54975	.83533	.65813	.5195	.1971	.8190	39
22	.54999	.83517	.65854	.5185	.1974	.8182	38
23	.55024	.83501	.65896	.5175	.1976	.8174	37
24	.55048	.83485	.65938	.5166	.1978	.8166	36
25	.55072	.83469	.65980	1.5156	1.1980	1.8158	35
26	.55097	.83453	.66021	.5147	.1983	.8150	34
27	.55121	.83437	.66063	.5137	.1985	.8142	33
28	.55145	.83421	.66105	.5127	.1987	.8134	32
29	.55169	.83405	.66147	.5118	.1990	.8126	31
30	.55194	.83388	.66188	1.5108	1.1992	1.8118	30
31	.55218	.83372	.66230	.5099	.1994	.8110	29
32	.55242	.83356	.66272	.5089	.1997	.8102	28
33	.55266	.83340	.66314	.5080	.1999	.8094	27
34	.55291	.83324	.66356	.5070	.2001	.8086	26
35	.55315	.83308	.66398	1.5061	1.2004	1.8078	25
36	.55339	.83292	.66440	.5051	.2006	.8070	24
37	.55363	.83276	.66482	.5042	.2008	.8062	23
38	.55388	.83260	.66524	.5032	.2010	.8054	22
39	.55412	.83244	.66566	.5023	.2013	.8047	21
40	.55436	.83228	.66608	1.5013	1.2015	1.8039	20
41	.55460	.83211	.66650	.5004	.2017	.8031	19
42	.55484	.83195	.66692	.4994	.2020	.8023	18
43	.55509	.83179	.66734	.4985	.2022	.8015	17
44	.55533	.83163	.66776	.4975	.2024	.8007	16
45	.55557	.83147	.66818	1.4966	1.2027	1.7999	15
46	.55581	.83131	.66860	.4957	.2029	.7992	14
47	.55605	.83115	.66902	.4947	.2031	.7984	13
48	.55629	.83098	.66944	.4938	.2034	.7976	12
49	.55654	.83082	.66986	.4928	.2036	.7968	11
50	.55678	.83066	.67028	1.4919	1.2039	1.7960	10
51	.55702	.83050	.67071	.4910	.2041	.7953	9
52	.55726	.83034	.67113	.4900	.2043	.7945	8
53	.55750	.83017	.67155	.4891	.2046	.7937	7
54	.55774	.83001	.67197	.4881	.2048	.7929	6
55	.55799	.82985	.67239	1.4872	1.2050	1.7921	5
56	.55823	.82969	.67282	.4863	.2053	.7914	4
57	.55847	.82952	.67324	.4853	.2055	.7906	3
58	.55871	.82936	.67366	.4844	.2057	.7898	2
59	.55895	.82920	.67408	.4835	.2060	.7891	1
60	.55919	.82904	.67451	1.4826	1.2062	1.7883	0
M	Cosine	Sine	Cotan.	Tan.	Cosec.	Secant	M

56°

34°

M	Sine	Cosine	Tan.	Cotan.	Secant	Cosec.	M
0	.55919	.82904	.67451	1.4826	1.2062	1.7883	60
1	.55943	.82887	.67493	.4816	.2064	.7875	59
2	.55967	.82871	.67535	.4807	.2067	.7867	58
3	.55992	.82855	.67578	.4798	.2069	.7860	57
4	.56016	.82839	.67620	.4788	.2072	.7852	56
5	.56040	.82822	.67663	1.4779	1.2074	1.7844	55
6	.56064	.82806	.67705	.4770	.2076	.7837	54
7	.56088	.82790	.67747	.4761	.2079	.7829	53
8	.56112	.82773	.67790	.4751	.2081	.7821	52
9	.56136	.82757	.67832	.4742	.2083	.7814	51
10	.56160	.82741	.67875	1.4733	1.2086	1.7806	50
11	.56184	.82724	.67917	.4724	.2088	.7798	49
12	.56208	.82708	.67960	.4714	.2091	.7791	48
13	.56232	.82692	.68002	.4705	.2093	.7783	47
14	.56256	.82675	.68045	.4696	.2095	.7776	46
15	.56280	.82659	.68087	1.4687	1.2098	1.7768	45
16	.56304	.82643	.68130	.4678	.2100	.7760	44
17	.56328	.82626	.68173	.4669	.2103	.7753	43
18	.56353	.82610	.68215	.4659	.2105	.7745	42
19	.56377	.82593	.68258	.4650	.2107	.7738	41
20	.56401	.82577	.68301	1.4641	1.2110	1.7730	40
21	.56425	.82561	.68343	.4632	.2112	.7723	39
22	.56449	.82544	.68386	.4623	.2115	.7715	38
23	.56473	.82528	.68429	.4614	.2117	.7708	37
24	.56497	.82511	.68471	.4605	.2119	.7700	36
25	.56521	.82495	.68514	1.4595	1.2122	1.7693	35
26	.56545	.82478	.68557	.4586	.2124	.7685	34
27	.56569	.82462	.68600	.4577	.2127	.7678	33
28	.56593	.82445	.68642	.4568	.2129	.7670	32
29	.56617	.82429	.68685	.4559	.2132	.7663	31
30	.56641	.82413	.68728	1.4550	1.2134	1.7655	30
31	.56664	.82396	.68771	.4541	.2136	.7648	29
32	.56688	.82380	.68814	.4532	.2139	.7640	28
33	.56712	.82363	.68857	.4523	.2141	.7633	27
34	.56736	.82347	.68899	.4514	.2144	.7625	26
35	.56760	.82330	.68942	1.4505	1.2146	1.7618	25
36	.56784	.82314	.68985	.4496	.2149	.7610	24
37	.56808	.82297	.69028	.4487	.2151	.7603	23
38	.56832	.82280	.69071	.4478	.2153	.7596	22
39	.56856	.82264	.69114	.4469	.2156	.7588	21
40	.56880	.82247	.69157	1.4460	1.2158	1.7581	20
41	.56904	.82231	.69200	.4451	.2161	.7573	19
42	.56928	.82214	.69243	.4442	.2163	.7566	18
43	.56952	.82198	.69286	.4433	.2166	.7559	17
44	.56976	.82181	.69329	.4424	.2168	.7551	16
45	.57000	.82165	.69372	1.4415	1.2171	1.7544	15
46	.57023	.82148	.69415	.4406	.2173	.7537	14
47	.57047	.82131	.69459	.4397	.2175	.7529	13
48	.57071	.82115	.69502	.4388	.2178	.7522	12
49	.57095	.82098	.69545	.4379	.2180	.7514	11
50	.57119	.82082	.69588	1.4370	1.2183	1.7507	10
51	.57143	.82065	.69631	.4361	.2185	.7500	9
52	.57167	.82048	.69674	.4352	.2188	.7493	8
53	.57191	.82032	.69718	.4343	.2190	.7485	7
54	.57214	.82015	.69761	.4335	.2193	.7478	6
55	.57238	.81998	.69804	1.4326	1.2195	1.7471	5
56	.57262	.81982	.69847	.4317	.2198	.7463	4
57	.57286	.81965	.69891	.4308	.2200	.7456	3
58	.57310	.81948	.69934	.4299	.2203	.7449	2
59	.57334	.81932	.69977	.4290	.2205	.7442	1
60	.57358	.81915	.70021	1.4281	1.2208	1.7434	0
M	Cosine	Sine	Cotan.	Tan.	Cosec.	Secant	M

55°

35°

M	Sine	Cosine	Tan.	Cotan.	Secant	Cosec.	M
0	.57358	.81915	.70021	1.4281	1.2208	1.7434	60
1	.57381	.81898	.70064	.4273	.2210	.7427	59
2	.57405	.81882	.70107	.4264	.2213	.7420	58
3	.57429	.81865	.70151	.4255	.2215	.7413	57
4	.57453	.81848	.70194	.4246	.2218	.7405	56
5	.57477	.81832	.70238	1.4237	1.2220	1.7398	55
6	.57500	.81815	.70281	.4228	.2223	.7391	54
7	.57524	.81798	.70325	4220	.2225	.7384	53
8	.57548	.81781	.70368	.4211	.2228	.7377	52
9	.57572	.81765	.70412	.4202	.2230	.7369	51
10	.57596	.81748	.70455	1.4193	1.2233	1.7362	50
11	.57619	.81731	.70499	.4185	.2235	.7355	49
12	.57643	.81714	.70542	.4176	.2238	.7348	48
13	.57667	.81698	.70586	.4167	.2240	.7341	47
14	.57691	.81681	.70629	.4158	.2243	.7334	46
15	.57714	.81664	.70673	1.4150	1.2245	1.7327	45
16	.57738	.81647	.70717	.4141	.2248	.7319	44
17	.57762	.81630	.70760	.4132	.2250	.7312	43
18	.57786	.81614	.70804	.4123	.2253	.7305	42
19	.57809	.81597	.70848	.4115	.2255	.7298	41
20	.57833	.81580	.70891	1.4106	1.2258	1.7291	40
21	.57857	.81563	.70935	.4097	.2260	.7284	39
22	.57881	.81546	.70979	.4089	.2263	.7277	38
23	.57904	.81530	.71022	.4080	.2265	.7270	37
24	.57928	.81513	.71066	.4071	.2268	.7263	36
25	.57952	.81496	.71110	1.4063	1.2270	1.7256	35
26	.57975	.81479	.71154	.4054	.2273	.7249	34
27	.57999	.81462	.71198	.4045	.2276	.7242	33
28	.58023	.81445	.71241	1.4037	.2278	.7234	32
29	.58047	.81428	.71285	.4028	.2281	.7227	31
30	.58070	.81411	.71329	1.4019	1.2283	1.7220	30
31	.58094	.81395	.71373	.4011	.2286	.7213	29
32	.58118	.81378	.71417	.4002	.2288	.7206	28
33	.58141	.81361	.71461	.3994	.2291	.7199	27
34	.58165	.81344	.71505	.3985	.2293	.7192	26
35	.58189	.81327	.71549	1.3976	1.2296	1.7185	25
36	.58212	.81310	.71593	.3968	.2298	.7178	24
37	.58236	.81293	.71637	.3959	1.2301	.7171	23
38	.58259	.81276	.71681	.3951	.2304	.7164	22
39	.58283	.81259	.71725	.3942	.2306	.7157	21
40	.58307	.81242	.71769	1.3933	1.2309	1.7151	20
41	.58330	.81225	.71813	.3925	.2311	.7144	19
42	.58354	.81208	.71857	.3916	.2314	.7137	18
43	.58378	.81191	.71901	.3908	.2316	.7130	17
44	.58401	.81174	.71945	.3899	.2319	.7123	16
45	.58425	.81157	.71990	1.3891	1.2322	1.7116	15
46	.58448	.81140	.72034	.3882	.2324	.7109	14
47	.58472	.81123	.72078	.3874	.2327	.7102	13
48	.58496	.81106	.72122	.3865	.2329	.7095	12
49	.58519	.81089	.72166	.3857	.2332	.7088	11
50	.58543	.81072	.72211	1.3848	1.2335	1.7081	10
51	.58566	.81055	.72255	.3840	.2337	.7075	9
52	.58590	.81038	.72299	1.3831	.2340	.7068	8
53	.58614	.81021	.72344	.3823	.2342	.7061	7
54	.58637	.81004	.72388	.3814	.2345	.7054	6
55	.58661	.80987	.72432	1.3806	1.2348	1.7047	5
56	.58684	.80970	.72477	.3797	.2350	.7040	4
57	.58708	.80953	.72521	.3789	.2353	.7033	3
58	.58731	.80936	.72565	.3781	.2355	.7027	2
59	.58755	.80919	.72610	.3772	.2358	.7020	1
60	.58778	.80902	.72654	1.3764	1.2361	1.7013	0
M	Cosine	Sine	Cotan.	Tan.	Cosec.	Secant	M

54°

36°

M	Sine	Cosine	Tan.	Cotan.	Secant	Cosec.	M
0	.58778	.80902	.72654	1.3764	1.2361	1.7013	60
1	.58802	.80885	.72699	.3755	.2363	.7006	59
2	.58825	.80867	.72743	.3747	.2366	.6999	58
3	.58849	.80850	.72788	.3738	.2368	.6993	57
4	.58873	.80833	.72832	1.3730	.2371	.6986	56
5	.58896	.80816	.72877	1.3722	1.2374	1.6979	55
6	.58920	.80799	.72921	.3713	.2376	.6972	54
7	.58943	.80782	.72966	.3705	.2379	.6965	53
8	.58967	.80765	.73010	.3697	.2382	.6959	52
9	.58990	.80747	.73055	.3688	.2384	.6952	51
10	.59014	.80730	.73100	1.3680	1.2387	1.6945	50
11	.59037	.80713	.73144	.3672	.2389	.6938	49
12	.59060	.80696	.73189	.3663	.2392	.6932	48
13	.59084	.80679	.73234	.3655	.2395	.6925	47
14	.59107	.80662	.73278	.3647	.2397	.6918	46
15	.59131	.80644	.73323	1.3638	1.2400	1.6912	45
16	.59154	.80627	.73368	.3630	.2403	.6905	44
17	.59178	.80610	.73412	.3622	.2405	.6898	43
18	.59201	.80593	.73457	.3613	.2408	.6891	42
19	.59225	.80576	.73502	.3605	.2411	.6885	41
20	.59248	.80558	.73547	1.3597	1.2413	1.6878	40
21	.59272	.80541	.73592	.3588	.2416	.6871	39
22	.59295	.80524	.73637	.3580	.2419	.6865	38
23	.59318	.80507	.73681	.3572	.2421	.6858	37
24	.59342	.80489	.73726	.3564	.2424	.6851	36
25	.59365	.80472	.73771	1.3555	1.2427	1.6845	35
26	.59389	.80455	.73816	.3547	.2429	1.6838	34
27	.59412	.80437	.73861	.3539	.2432	.6831	33
28	.59435	.80420	.73906	.3531	.2435	.6825	32
29	.59459	.80403	.73951	.3522	.2437	.6818	31
30	.59482	.80386	.73996	1.3514	1.2440	1.6812	30
31	.59506	.80368	.74041	.3506	.2443	.6805	29
32	.59529	.80351	.74086	.3498	.2445	.6798	28
33	.59552	.80334	.74131	.3489	.2448	.6792	27
34	.59576	.80316	.74176	.3481	.2451	.6785	26
35	.59599	.80299	.74221	1.3473	1.2453	1.6779	25
36	.59622	.80282	.74266	.3465	.2456	.6772	24
37	.59646	.80264	.74312	.3457	.2459	.6766	23
38	.59669	.80247	.74357	.3449	.2461	.6759	22
39	.59692	.80230	.74402	.3440	.2464	.6752	21
40	.59716	.80212	.74447	1.3432	1.2467	1.6746	20
41	.59739	.80195	.74492	.3424	.2470	.6739	19
42	.59762	.80177	.74538	.3416	.2472	.6733	18
43	.59786	.80160	.74583	.3408	.2475	.6726	17
44	.59809	.80143	.74628	.3400	.2478	.6720	16
45	.59832	.80125	.74673	1.3392	1.2480	1.6713	15
46	.59856	.80108	.74719	1.3383	.2483	.6707	14
47	.59879	.80090	.74764	.3375	.2486	.6700	13
48	.59902	.80073	.74809	.3367	.2488	.6694	12
49	.59926	.80056	.74855	.3359	.2491	.6687	11
50	.59949	.80038	.74900	1.3351	1.2494	1.6681	10
51	.59972	.80021	.74946	.3343	.2497	.6674	9
52	.59995	.80003	.74991	.3335	.2499	.6668	8
53	.60019	.79986	.75037	.3327	.2502	.6661	7
54	.60042	.79968	.75082	.3319	.2505	.6655	6
55	.60065	.79951	.75128	1.3311	1.2508	1.6648	5
56	.60088	.79933	.75173	.3303	.2510	.6642	4
57	.60112	.79916	.75219	.3294	.2513	.6636	3
58	.60135	.79898	.75264	.3286	.2516	.6629	2
59	.60158	.79881	.75310	.3278	.2519	.6623	1
60	.60181	.79863	.75355	1.3270	1.2521	1.6616	0
M	Cosine	Sine	Cotan.	Tan.	Cosec.	Secant	M

53°

37°

M	Sine	Cosine	Tan.	Cotan.	Secant	Cosec.	M
0	.60181	.79863	.75355	1.3270	1.2521	1.6616	60
1	.60205	.79846	.75401	.3262	.2524	.6610	59
2	.60228	.79828	.75447	.3254	.2527	.6603	58
3	.60251	.79811	.75492	.3246	.2530	.6597	57
4	.60274	.79793	.75538	.3238	.2532	.6591	56
5	.60298	.79776	.75584	1.3230	1.2535	1.6584	55
6	.60320	.79758	.75629	.3222	.2538	.6578	54
7	.60344	.79741	.75675	.3214	.2541	.6572	53
8	.60367	.79723	.75721	.3206	.2543	.6565	52
9	.60390	.79706	.75767	.3198	.2546	.6559	51
10	.60413	.79688	.75812	1.3190	1.2549	1.6552	50
11	.60437	.79670	.75858	.3182	.2552	.6546	49
12	.60460	.79653	.75904	.3174	.2554	.6540	48
13	.60483	.79635	.75950	.3166	.2557	.6533	47
14	.60506	.79618	.75996	.3159	.2560	.6527	46
15	.60529	.79600	.76042	1.3151	1.2563	1.6521	45
16	.60552	.79582	.76088	.3143	.2565	.6514	44
17	.60576	.79565	.76134	.3135	.2568	.6508	43
18	.60599	.79547	.76179	.3127	.2571	.6502	42
19	.60622	.79530	.76225	.3119	.2574	.6496	41
20	.60645	.79512	.76271	1.3111	1.2577	1.6489	40
21	.60668	.79494	.76317	.3103	.2579	.6483	39
22	.60691	.79477	.76364	.3095	.2582	.6477	38
23	.60714	.79459	.76410	.3087	.2585	.6470	37
24	.60737	.79441	.76456	.3079	.2588	.6464	36
25	.60761	.79424	.76502	1.3071	1.2591	1.6458	35
26	.60784	.79406	.76548	.3064	.2593	.6452	34
27	.60807	.79388	.76594	.3056	.2596	.6445	33
28	.60830	.79371	.76640	.3048	.2599	.6439	32
29	.60853	.79353	.76686	.3040	.2602	.6433	31
30	.60876	.79335	.76733	1.3032	1.2605	1.6427	30
31	.60899	.79318	.76779	.3024	.2607	.6420	29
32	.60922	.79300	.76825	.3016	.2610	.6414	28
33	.60945	.79282	.76871	.3009	.2613	.6408	27
34	.60968	.79264	.76918	.3001	.2616	.6402	26
35	.60991	.79247	.76964	1.2993	1.2619	1.6396	25
36	.61014	.79229	.77010	.2985	.2622	1.6389	24
37	.61037	.79211	.77057	.2977	.2624	1.6383	23
38	.61061	.79193	.77103	.2970	.2627	.6377	22
39	.61084	.79176	.77149	.2962	.2630	.6371	21
40	.61107	.79158	.77196	1.2954	1.2633	1.6365	20
41	.61130	.79140	.77242	.2946	.2636	.6359	19
42	.61153	.79122	.77289	.2938	.2639	.6352	18
43	.61176	.79104	.77335	.2931	.2641	.6346	17
44	.61199	.79087	.77382	.2923	.2644	.6340	16
45	.61222	.79069	.77428	1.2915	1.2647	1.6334	15
46	.61245	.79051	.77475	.2907	.2650	.6328	14
47	.61268	.79033	.77521	.2900	.2653	.6322	13
48	.61290	.79015	.77568	.2892	.2656	.6316	12
49	.61314	.78998	.77614	.2884	.2659	.6309	11
50	.61337	.78980	.77661	1.2876	1.2661	1.6303	10
51	.61360	.78962	.77708	.2869	.2664	.6297	9
52	.61383	.78944	.77754	.2861	.2667	.6291	8
53	.61405	.78926	.77801	.2853	.2670	.6285	7
54	.61428	.78908	.77848	.2845	.2673	.6279	6
55	.61451	.78890	.77895	1.2838	1.2676	1.6273	5
56	.61474	.78873	.77941	.2830	.2679	.6267	4
57	.61497	.78855	.77988	.2822	.2681	.6261	3
58	.61520	.78837	.78035	.2815	.2684	.6255	2
59	.61543	.78819	.78082	.2807	.2687	.6249	1
60	.61566	.78801	.78128	1.2799	1.2690	1.6243	0
M	Cosine	Sine	Cotan.	Tan.	Cosec.	Secant	M

52°

38°

M	Sine	Cosine	Tan.	Cotan.	Secant	Cosec.	M
0	.61566	.78801	.78128	1.2799	1.2690	1.6243	60
1	.61589	.78783	.78175	.2792	.2693	.6237	59
2	.61612	.78765	.78222	.2784	.2696	.6231	58
3	.61635	.78747	.78269	.2776	.2699	.6224	57
4	.61658	.78729	.78316	.2769	.2702	.6218	56
5	.61681	.78711	.78363	1.2761	1.2705	1.6212	55
6	.61703	.78693	.78410	.2753	.2707	.6206	54
7	.61726	.78675	.78457	.2746	.2710	.6200	53
8	.61749	.78657	.78504	.2738	.2713	.6194	52
9	.61772	.78640	.78551	.2730	.2716	.6188	51
10	.61795	.78622	.78598	1.2723	1.2719	1.6182	50
11	.61818	.78604	.78645	.2715	.2722	.6176	49
12	.61841	.78586	.78692	.2708	.2725	.6170	48
13	.61864	.78568	.78739	.2700	.2728	.6164	47
14	.61886	.78550	.78786	.2692	.2731	.6159	46
15	.61909	.78532	.78834	1.2685	1.2734	1.6153	45
16	.61932	.78514	.78881	.2677	.2737	.6147	44
17	.61955	.78496	.78928	.2670	.2739	.6141	43
18	.61978	.78478	.78975	.2662	.2742	.6135	42
19	.62001	.78460	.79022	.2655	.2745	.6129	41
20	.62023	.78441	.79070	1.2647	1.2748	1.6123	40
21	.62046	.78423	.79117	.2639	.2751	.6117	39
22	.62069	.78405	.79164	.2632	.2754	.6111	38
23	.62092	.78387	.79212	.2624	.2757	.6105	37
24	.62115	.78369	.79259	.2617	.2760	.6099	36
25	.62137	.78351	.79306	1.2609	1.2763	1.6093	35
26	.62160	.78333	.79354	.2602	.2766	.6087	34
27	.62183	.78315	.79401	.2594	.2769	.6081	33
28	.62206	.78297	.79449	.2587	.2772	.6077	32
29	.62229	.78279	.79496	.2579	.2775	.6070	31
30	.62251	.78261	.79543	1.2572	1.2778	1.6064	30
31	.62274	.78243	.79591	.2564	.2781	.6058	29
32	.62297	.78224	.79639	.2557	.2784	.6052	28
33	.62320	.78206	.79686	.2549	.2787	.6046	27
34	.62342	.78188	.79734	.2542	.2790	.6040	26
35	.62365	.78170	.79781	1.2534	1.2793	1.6034	25
36	.62388	.78152	.79829	.2527	.2795	.6029	24
37	.62411	.78134	.79876	.2519	.2798	.6023	23
38	.62433	.78116	.79924	.2512	.2801	.6017	22
39	.62456	.78097	.79972	.2504	.2804	.6011	21
40	.62479	.78079	.80020	1.2497	1.2807	1.6005	20
41	.62501	.78061	.80067	.2489	.2810	.6000	19
42	.62524	.78043	.80115	.2482	.2813	.5994	18
43	.62547	.78025	.80163	.2475	.2816	.5988	17
44	.62570	.78007	.80211	.2467	.2819	.5982	16
45	.62592	.77988	.80258	1.2460	1.2822	1.5976	15
46	.62615	.77970	.80306	.2452	.2825	.5971	14
47	.62638	.77952	.80354	.2445	.2828	.5965	13
48	.62660	.77934	.80402	.2437	.2831	.5959	12
49	.62683	.77915	.80450	.2430	.2834	.5953	11
50	.62706	.77897	.80498	1.2423	1.2837	1.5947	10
51	.62728	.77879	.80546	.2415	.2840	.5942	9
52	.62751	.77861	.80594	.2408	.2843	.5936	8
53	.62774	.77842	.80642	.2400	.2846	.5930	7
54	.62796	.77824	.80690	.2393	.2849	.5924	6
55	.62819	.77806	.80738	1.2386	1.2852	1.5919	5
56	.62841	.77788	.80786	.2378	.2855	.5913	4
57	.62864	.77769	.80834	.2371	.2858	.5907	3
58	.62887	.77751	.80882	.2364	.2861	.5901	2
59	.62909	.77733	.80930	.2356	.2864	.5896	1
60	.62932	.77715	.80978	1.2349	1.2867	1.5890	0
M	Cosine	Sine	Cotan.	Tan.	Cosec.	Secant	M

51°

39°

M	Sine	Cosine	Tan.	Cotan.	Secant	Cosec.	M
0	.62932	.77715	.80978	1.2349	1.2867	1.5890	60
1	.62955	.77696	.81026	.2342	.2871	.5884	59
2	.62977	.77678	.81075	.2334	.2874	.5879	58
3	.63000	.77660	.81123	.2327	.2877	.5873	57
4	.63022	.77641	.81171	.2320	.2880	.5867	56
5	.63045	.77623	.81219	1.2312	1.2883	1.5862	55
6	.63067	.77605	.81268	.2305	.2886	.5856	54
7	.63090	.77586	.81316	.2297	.2889	.5850	53
8	.63113	.77568	.81364	.2290	.2892	.5845	52
9	.63135	.77549	.81413	.2283	.2895	.5839	51
10	.63158	.77531	.81461	1.2276	1.2898	1.5833	50
11	.63180	.77513	.81509	.2268	.2901	.5828	49
12	.63203	.77494	.81558	.2261	.2904	.5822	48
13	.63225	.77476	.81606	.2254	,2907	.5816	47
14	.63248	.77458	.81655	.2247	.2910	.5811	46
15	.63270	.77439	.81703	1.2239	1.2913	1.5805	45
16	.63293	.77421	.81752	.2232	.2916	.5799	44
17	.63315	.77402	.81800	.2225	.2919	.5794	43
18	.63338	.77384	.81849	.2218	.2922	.5788	42
19	.63360	.77365	.81898	.2210	.2926	.5783	41
20	.63383	.77347	.81946	1.2203	1.2929	1.5777	40
21	.63405	.77329	.81995	.2196	.2932	.5771	39
22	.63428	.77310	.82043	.2189	.2935	.5766	38
23	.63450	.77292	.82092	.2181	.2938	.5760	37
24	.63473	.77273	.82141	.2174	.2941	.5755	36
25	.63495	.77255	.82190	1.2167	1.2944	1.5749	35
26	.63518	.77236	.82238	.2160	.2947	.5743	34
27	.63540	.77218	.82287	.2152	.2950	.5738	33
28	.63563	.77199	.82336	.2145	.2953	.5732	32
29	.63585	.77181	.82385	.2138	.2956	.5727	31
30	.63608	.77162	.82434	1.2131	1.2960	1.5721	30
31	.63630	.77144	.82482	.2124	.2963	.5716	29
32	.63653	.77125	.82531	.2117	.2966	.5710	28
33	.63675	.77107	.82580	.2109	.2969	.5705	27
34	.63697	.77088	.82629	.2102	.2972	.5699	26
35	.63720	.77070	.82678	1.2095	1.2975	1.5694	25
36	.63742	.77051	.82727	.2088	.2978	.5688	24
37	.63765	.77033	.82776	.2081	.2981	.5683	23
38	.63787	.77014	.82825	.2074	.2985	.5677	22
39	.63810	.76996	.82874	.2066	.2988	.5672	21
40	.63832	.76977	.82923	1.2059	1.2991	1.5666	20
41	.63854	.76958	.82972	.2052	.2994	.5661	19
42	.63877	.76940	.83022	.2045	.2997	.5655	18
43	.63899	.76921	.83071	.2038	.3000	.5650	17
44	.63921	.76903	.83120	.2031	.3003	.5644	16
45	.63944	.76884	.83169	1.2024	1.3006	1.5639	15
46	.63966	.76865	.83218	.2016	.3010	.5633	14
47	.63989	.76847	.83267	.2009	.3013	.5628	13
48	.64011	.76828	.83317	.2002	.3016	.5622	12
49	.64033	.76810	.83366	.1995	.3019	.5617	11
50	.64056	.76791	.83415	1.1988	1.3022	1.5611	10
51	.64078	.76772	.83465	.1981	.3025	.5606	9
52	.64100	.76754	.83514	.1974	.3029	.5600	8
53	.64123	.76735	.83563	.1967	.3032	.5595	7
54	.64145	.76716	.83613	.1960	.3035	.5590	6
55	.64167	.76698	.83662	1.1953	1.3038	1.5584	5
56	.64189	.76679	.83712	.1946	.3041	.5579	4
57	.64212	.76660	.83761	.1939	.3044	.5573	3
58	.64234	.76642	.83811	.1932	.3048	.5568	2
59	.64256	.76623	.83860	.1924	.3051	.5563	1
60	.64279	.76604	.83910	1.1917	1.3054	1.5557	0
M	Cosine	Sine	Cotan.	Tan.	Cosec.	Secant	M

50°

40°

M	Sine	Cosine	Tan.	Cotan.	Secant	Cosec.	M
0	.64279	.76604	.83910	1.1917	1.3054	1.5557	60
1	.64301	.76586	.83959	.1910	.3057	.5552	59
2	.64323	.76567	.84009	.1903	.3060	.5546	58
3	.64345	.76548	.84059	.1896	.3064	.5541	57
4	.64368	.76530	.84108	.1889	.3067	.5536	56
5	.64390	.76511	.84158	1.1882	1.3070	1.5530	55
6	.64412	.76492	.84208	.1875	.3073	.5525	54
7	.64435	.76473	.84257	.1868	.3076	.5520	53
8	.64457	.76455	.84307	.1861	.3080	.5514	52
9	.64479	.76436	.84357	.1854	.3083	.5509	51
10	.64501	.76417	.84407	1.1847	1.3086	1.5503	50
11	.64523	.76398	.84457	.1840	.3089	.5498	49
12	.64546	.76380	.84506	.1833	.3092	.5493	48
13	.64568	.76361	.84556	.1826	.3096	.5487	47
14	.64590	.76342	.84606	.1819	.3099	.5482	46
15	.64612	.76323	.84656	1.1812	1.3102	1.5477	45
16	.64635	.76304	.84706	.1805	.3105	.5471	44
17	.64657	.76286	.84756	.1798	.3109	.5466	43
18	.64679	.76267	.84806	.1791	.3112	.5461	42
19	.64701	.76248	.84856	.1785	.3115	.5456	41
20	.64723	.76229	.84906	1.1778	1.3118	1.5450	40
21	.64745	.76210	.84956	.1771	.3121	.5445	39
22	.64768	.76191	.85006	.1764	.3125	.5440	38
23	.64790	.76173	.85056	.1757	.3128	.5434	37
24	.64812	.76154	.85107	.1750	.3131	.5429	36
25	.64834	.76135	.85157	1.1743	1.3134	1.5424	35
26	.64856	.76116	.85207	.1736	.3138	.5419	34
27	.64878	.76097	.85257	.1729	.3141	.5413	33
28	.64900	.76078	.85307	.1722	.3144	.5408	32
29	.64923	.76059	.85358	.1715	.3148	.5403	31
30	.64945	.76041	.85408	1.1708	1.3151	1.5398	30
31	.64967	.76022	.85458	.1702	.3154	.5392	29
32	.64989	.76003	.85509	.1695	.3157	.5387	28
33	.65011	.75984	.85559	.1688	.3161	.5382	27
34	.65033	.75965	.85609	.1681	.3164	.5377	26
35	.65055	.75946	.85660	1.1674	1.3167	1.5371	25
36	.65077	.75927	.85710	.1667	.3170	.5366	24
37	.65100	.75908	.85761	.1660	.3174	.5361	23
38	.65121	.75889	.85811	.1653	.3177	.5356	22
39	.65144	.75870	.85862	.1647	.3180	.5351	21
40	.65166	.75851	.85912	1.1640	1.3184	1.5345	20
41	.65188	.75832	.85963	.1633	.3187	.5340	19
42	.65210	.75813	.86013	.1626	.3190	.5335	18
43	.65232	.75794	.86064	.1619	.3193	.5330	17
44	.65254	.75775	.86115	.1612	.3197	.5325	16
45	.65276	.75756	.86165	1.1605	1.3200	1.5319	15
46	.65298	.75737	.86216	.1599	.3203	.5314	14
47	.65320	.75718	.86267	.1592	.3207	.5309	13
48	.65342	.75700	.86318	.1585	.3210	.5304	12
49	.65364	.75680	.86368	.1578	.3213	.5290	11
50	.65386	.75661	.86419	1.1571	1.3217	1.5294	10
51	.65408	.75642	.86470	.1565	.3220	.5289	9
52	.65430	.75623	.86521	.1558	.3223	.5283	8
53	.65452	.75604	.86572	.1551	.3227	.5278	7
54	.65474	.75585	.86623	.1544	.3230	.5273	6
55	.65496	.75566	.86674	1.1537	1.3233	1.5268	5
56	.65518	.75547	.86725	.1531	.3237	.5263	4
57	.65540	.75528	.86775	.1524	.3240	.5258	3
58	.65562	.75509	.86826	.1517	.3243	.5253	2
59	.65584	.75490	.86878	.1510	.3247	.5248	1
60	.65606	.75471	.86929	1.1504	1.3250	1.5242	0
M	Cosine	Sine	Cotan.	Tan.	Cosec.	Secant	M

49°

41°

M	Sine	Cosine	Tan.	Cotan.	Secant	Cosec.	M
0	.65606	.75471	.86929	1.1504	1.3250	1.5242	60
1	.65628	.75452	.86980	.1497	.3253	.5237	59
2	.65650	.75433	.87031	.1490	.3257	.5232	58
3	.65672	.75414	.87082	.1483	.3260	.5227	57
4	.65694	.75394	.87133	.1477	.3263	.5222	56
5	.65716	.75375	.87184	1.1470	1.3267	1.5217	55
6	.65737	.75356	.87235	.1463	.3270	.5212	54
7	.65759	.75337	.87287	.1456	.3274	.5207	53
8	.65781	.75318	.87338	.1450	.3277	.5202	52
9	.65803	.75299	.87389	.1443	.3280	.5197	51
10	.65825	.75280	.87441	1.1436	1.3284	1.5192	50
11	.65847	.75261	.87492	.1430	.3287	.5187	49
12	.65869	.75241	.87543	.1423	.3290	.5182	48
13	.65891	.75222	.87595	.1416	.3294	.5177	47
14	.65913	.75203	.87646	.1409	.3297	.5171	46
15	.65934	.75184	.87698	1.1403	1.3301	1.5166	45
16	.65956	.75165	.87749	.1396	.3304	.5161	44
17	.65978	.75146	.87801	.1389	.3307	.5156	43
18	.66000	.75126	.87852	.1383	.3311	.5151	42
19	.66022	.75107	.87904	.1376	.3314	.5146	41
20	.66044	.75088	.87955	1.1369	1.3318	1.5141	40
21	.66066	.75069	.88007	.1363	.3321	.5136	39
22	.66087	.75049	.88058	.1356	.3324	.5131	38
23	.66109	.75030	.88110	.1349	.3328	.5126	37
24	.66131	.75011	.88162	.1343	.3331	.5121	36
25	.66153	.74992	.88213	1.1336	1.3335	1.5116	35
26	.66175	.74973	.88265	.1329	.3338	.5111	34
27	.66197	.74953	.88317	.1323	.3342	.5106	33
28	.66218	.74934	.88369	.1316	.3345	.5101	32
29	.66240	.74915	.88421	.1309	.3348	.5096	31
30	.66262	.74895	.88472	1.1303	1.3352	1.5092	30
31	.66284	.74876	.88524	.1296	.3355	.5087	29
32	.66305	.74857	.88576	.1290	.3359	.5082	28
33	.66327	.74838	.88628	.1283	.3362	.5077	27
34	.66349	.74818	.88680	.1276	.3366	.5072	26
35	.66371	.74799	.88732	1.1270	1.3369	1.5067	25
36	.66393	.74780	.88784	.1263	.3372	.5062	24
37	.66414	.74760	.88836	.1257	.3376	.5057	23
38	.66436	.74741	.88888	.1250	.3379	.5052	22
39	.66458	.74722	.88940	.1243	.3383	.5047	21
40	.66479	.74702	.88992	1.1237	1.3386	1.5042	20
41	.66501	.74683	.89044	.1230	.3390	.5037	19
42	.66523	.74664	.89097	.1224	.3393	.5032	18
43	.66545	.74644	.89149	.1217	.3397	.5027	17
44	.66566	.74625	.89201	.1211	.3400	.5022	16
45	.66588	.74606	.89253	1.1204	1.3404	1.5018	15
46	.66610	.74586	.89306	.1197	.3407	.5013	14
47	.66631	.74567	.89358	.1191	.3411	.5008	13
48	.66653	.74548	.89410	.1184	.3414	.5003	12
49	.66675	.74528	.89463	.1178	.3418	.4998	11
50	.66697	.74509	.89515	1.1171	1.3421	1.4993	10
51	.66718	.74489	.89567	.1165	.3425	.4988	9
52	.66740	.74470	.89620	.1158	.3428	.4983	8
53	.66762	.74450	.89672	.1152	.3432	.4979	7
54	.66783	.74431	.89725	.1145	.3435	.4974	6
55	.66805	.74412	.89777	1.1139	1.3439	1.4969	5
56	.66826	.74392	.89830	.1132	.3442	.4964	4
57	.66848	.74373	.89882	.1126	.3446	.4959	3
58	.66870	.74353	.89935	.1119	.3449	.4954	2
59	.66891	.74334	.89988	.1113	.3453	.4949	1
60	.66913	.74314	.90040	1.1106	1.3456	1.4945	0
M	Cosine	Sine	Cotan.	Tan.	Cosec.	Secant	M

48°

42°

M	Sine	Cosine	Tan.	Cotan.	Secant	Cosec.	M
0	.66913	.74314	.90040	1.1106	1.3456	1.4945	60
1	.66935	.74295	.90093	.1100	.3460	.4940	59
2	.66956	.74275	.90146	.1093	.3463	.4935	58
3	.66978	.74256	.90198	.1086	.3467	.4930	57
4	.66999	.74236	.90251	.1080	.3470	.4925	56
5	.67021	.74217	.90304	1.1074	1.3474	1.4921	55
6	.67043	.74197	.90357	.1067	.3477	.4916	54
7	.67064	.74178	.90410	.1061	.3481	.4911	53
8	.67086	.74158	.90463	.1054	.3485	.4906	52
9	.67107	.74139	.90515	.1048	.3488	.4901	51
10	.67129	.74119	.90568	1.1041	1.3492	1.4897	50
11	.67150	.74100	.90621	.1035	.3495	.4892	49
12	.67172	.74080	.90674	.1028	.3499	.4887	48
13	.67194	.74061	.90727	.1022	.3502	.4882	47
14	.67215	.74041	.90780	.1015	.3506	.4877	46
15	.67237	.74022	.90834	1.1009	1.3509	1.4873	45
16	.67258	.74002	.90887	.1003	.3513	.4868	44
17	.67280	.73983	.90940	.0996	.3517	.4863	43
18	.67301	.73963	.90993	.0990	.3520	.4858	42
19	.67323	.73943	.91046	.0983	.3524	.4854	41
20	.67344	.73924	.91099	1.0977	1.3527	1.4849	40
21	.67366	.73904	.91153	.0971	.3531	.4844	39
22	.67387	.73885	.91206	.0964	.3534	.4839	38
23	.67409	.73865	.91259	.0958	.3538	.4835	37
24	.67430	.73845	.91312	.0951	.3542	.4830	36
25	.67452	.73826	.91366	1.0945	1.3545	1.4825	35
26	.67473	.73806	.91419	.0939	.3549	.4821	34
27	.67495	.73787	.91473	.0932	.3552	.4816	33
28	.67516	.73767	.91526	.0926	.3556	.4811	32
29	.67537	.73747	.91580	.0919	.3560	.4806	31
30	.67559	.73728	.91633	1.0913	1.3563	1.4802	30
31	.67580	.73708	.91687	.0907	.3567	.4797	29
32	.67602	.73688	.91740	.0900	.3571	.4792	28
33	.67623	.73669	.91794	.0894	.3574	.4788	27
34	.67645	.73649	.91847	.0888	.3578	.4783	26
35	.67666	.73629	.91901	1.0881	1.3581	1.4778	25
36	.67688	.73610	.91955	.0875	.3585	.4774	24
37	.67709	.73590	.92008	.0868	.3589	.4769	23
38	.67730	.73570	.92062	.0862	.3592	.4764	22
39	.67752	.73551	.92116	.0856	.3596	.4760	21
40	.67773	.73531	.92170	1.0849	1.3600	1.4755	20
41	.67794	.73511	.92223	.0843	.3603	.4750	19
42	.67816	.73491	.92277	.0837	.3607	.4746	18
43	.67837	.73472	.92331	.0830	.3611	.4741	17
44	.67859	.73452	.92385	.0824	.3614	.4736	16
45	.67880	.73432	.92439	1.0818	1.3618	1.4732	15
46	.67901	.73412	.92493	.0812	.3622	.4727	14
47	.67923	.73393	.92547	.0805	.3625	.4723	13
48	.67944	.73373	.92601	.0799	.3629	.4718	12
49	.67965	.73353	.92655	.0793	.3633	.4713	11
50	.67987	.73333	.92709	1.0786	1.3636	1.4709	10
51	.68008	.73314	.92763	.0780	.3640	.4704	9
52	.68029	.73294	.92817	.0774	.3644	.4699	8
53	.68051	.73274	.92871	.0767	.3647	.4695	7
54	.68072	.73254	.92926	.0761	.3651	.4690	6
55	.68093	.73234	.92980	1.0755	1.3655	1.4686	5
56	.68115	.73215	.93034	.0749	.3658	.4681	4
57	.68136	.73195	.93088	.0742	.3662	.4676	3
58	.68157	.73175	.93143	.0736	.3666	.4672	2
59	.68178	.73155	.93197	.0730	.3669	.4667	1
60	.68200	.73135	.93251	1.0724	1.3673	1.4663	0
M	Cosine	Sine	Cotan.	Tan.	Cosec.	Secant	M

47°

43°

M	Sine	Cosine	Tan.	Cotan.	Secant	Cosec.	M
0	.68200	.73135	.93251	1.0724	1.3673	1.4663	60
1	.68221	.73115	.93306	.0717	.3677	.4658	59
2	.68242	.73096	.93360	.0711	.3681	.4654	58
3	.68264	.73076	.93415	.0705	.3684	.4649	57
4	.68285	.73056	.93469	.0699	.3688	.4644	56
5	.68306	.73036	.93524	1.0692	1.3692	1.4640	55
6	.68327	.73016	.93578	.0686	.3695	.4635	54
7	.68349	.72996	.93633	.0680	.3699	.4631	53
8	.68370	.72976	.93687	.0674	.3703	.4626	52
9	.68391	.72956	.93742	.0667	.3707	.4622	51
10	.68412	.72937	.93797	1.0661	1.3710	1.4617	50
11	.68433	.72917	.93851	.0655	.3714	.4613	49
12	.68455	.72897	.93906	.0649	.3718	.4608	48
13	.68476	.72877	.93961	.0643	.3722	.4604	47
14	.68497	.72857	.94016	.0636	.3725	.4599	46
15	.68518	.72837	.94071	1.0630	1.3729	1.4595	45
16	68539	.72817	.94125	.0624	.3733	.4590	44
17	.68561	.72797	.94180	.0618	.3737	.4586	43
18	.68582	.72777	.94235	.0612	.3740	.4581	42
19	.68603	.72757	.94290	.0605	.3744	.4577	41
20	.68624	.72737	.94345	1.0599	1.3748	1.4572	40
21	.68645	.72717	.94400	.0593	.3752	.4568	39
22	.68666	.72697	.94455	.0587	.3756	.4563	38
23	.68688	.72677	.94510	.0581	.3759	.4559	37
24	.68709	.72657	.94565	.0575	.3763	.4554	36
25	.68730	.72637	.94620	1.0568	1.3767	1.4550	35
26	.68751	.72617	.94675	.0562	.3771	.4545	34
27	.68772	.72597	.94731	.0556	.3774	.4541	33
28	.68793	.72577	.94786	.0550	.3778	.4536	32
29	.68814	.72557	.94841	.0544	.3782	.4532	31
30	.68835	.72537	.94896	1.0538	1.3786	1.4527	30
31	.68856	.72517	.94952	.0532	.3790	.4523	29
32	.68878	.72497	.95007	.0525	.3794	.4518	28
33	.68899	.72477	.95062	.0519	.3797	.4514	27
34	.68920	.72457	.95118	.0513	.3801	.4510	26
35	.68941	.72437	.95173	1.0507	1.3805	1.4505	25
36	.68962	.72417	.95229	.0501	.3809	.4501	24
37	.68983	.72397	.95284	.0495	.3813	.4496	23
38	.69004	.72377	.95340	.0489	.3816	.4492	22
39	.69025	.72357	.95395	.0483	.3820	.4487	21
40	.69046	.72337	.95451	1.0476	1.3824	1.4483	20
41	.69067	.72317	.95506	.0470	.3828	.4479	19
42	.69088	.72297	.95562	.0464	.3832	.4474	18
43	.69109	.72277	.95618	.0458	.3836	.4470	17
44	.69130	.72256	.95673	.0452	.3839	.4465	16
45	.69151	.72236	.95729	1.0446	1.3843	1.4461	15
46	.69172	.72216	.95785	.0440	.3847	.4457	14
47	.69193	.72196	.95841	.0434	.3851	.4452	13
48	.69214	.72176	.95896	.0428	.3855	.4448	12
49	.69235	.72156	.95952	.0422	.3859	.4443	11
50	.69256	.72136	.96008	1.0416	1.3863	1.4439	10
51	.69277	.72115	.96064	.0410	.3867	.4435	9
52	.69298	.72095	.96120	.0404	.3870	.4430	8
53	.69319	.72075	.96176	.0397	.3874	.4426	7
54	.69340	.72055	.96232	.0391	.3878	.4422	6
55	.69361	.72035	.96288	1.0385	1.3882	1.4417	5
56	.69382	.72015	.96344	.0379	.3886	.4413	4
57	.69403	.71994	.96400	.0373	.3890	.4408	3
58	.69424	.71974	.96456	.0367	.3894	.4404	2
59	.69445	.71954	.96513	.0361	.3898	.4400	1
60	.69466	.71934	.96569	1.0355	1.3902	1.4395	0
M	Cosine	Sine	Cotan.	Tan.	Cosec.	Secant	M

46°

44°

M	Sine	Cosine	Tan.	Cotan.	Secant	Cosec.	M
0	.69466	.71934	.96569	1.0355	1.3902	1.4395	60
1	.69487	.71914	.96625	.0349	.3905	.4391	59
2	.69508	.71893	.96681	.0343	.3909	.4387	58
3	.69528	.71873	.96738	.0337	.3913	.4382	57
4	.69549	.71853	.96794	.0331	.3917	.4378	56
5	.69570	.71833	.96850	1.0325	1.3921	1.4374	55
6	.69591	.71813	.96907	.0319	.3925	.4370	54
7	.69612	.71792	.96963	.0313	.3929	.4365	53
8	.69633	.71772	.97020	.0307	.3933	.4361	52
9	.69654	.71752	.97076	.0301	.3937	.4357	51
10	.69675	.71732	.97133	1.0295	1.3941	1.4352	50
11	.69696	.71711	.97189	.0289	.3945	.4348	49
12	.69716	.71691	.97246	.0283	.3949	.4344	48
13	.69737	.71671	.97302	.0277	.3953	.4339	47
14	.69758	.71650	.97359	.0271	.3957	.4335	46
15	.69779	.71630	.97416	1.0265	1.3960	1.4331	45
16	.69800	.71610	.97472	.0259	.3964	.4327	44
17	.69821	.71589	.97529	.0253	.3968	.4322	43
18	.69841	.71569	.97586	.0247	.3972	.4318	42
19	.69862	.71549	.97643	.0241	.3976	.4314	41
20	.69883	.71529	.97700	1.0235	1.3980	1.4310	40
21	.69904	.71508	.97756	.0229	.3984	.4305	39
22	.69925	.71488	.97813	.0223	.3988	.4301	38
23	.69945	.71468	.97870	.0218	.3992	.4297	37
24	.69966	.71447	.97927	.0212	.3996	.4292	36
25	.69987	.71427	.97984	1.0206	1.4000	1.4288	35
26	.70008	.71406	.98041	.0200	.4004	.4284	34
27	.70029	.71386	.98098	.0194	.4008	.4280	33
28	.70049	.71366	.98155	.0188	.4012	.4276	32
29	.70070	.71345	.98212	.0182	.4016	.4271	31
30	.70091	.71325	.98270	1.0176	1.4020	1.4267	30
31	.70112	.71305	.98327	.0170	.4024	.4263	29
32	.70132	.71284	.98384	.0164	.4028	.4259	28
33	.70153	.71264	.98441	.0158	.4032	.4254	27
34	.70174	.71243	.98499	.0152	.4036	.4250	26
35	.70194	.71223	.98556	1.0146	1.4040	1.4246	25
36	.70215	.71203	.98613	.0141	.4044	.4242	24
37	.70236	.71182	.98671	.0135	.4048	.4238	23
38	.70257	.71162	.98728	.0129	.4052	.4233	22
39	.70277	.71141	.98786	.0123	.4056	.4229	21
40	.70298	.71121	.98843	1.0117	1.4060	1.4225	20
41	.70319	.71100	.98901	.0111	.4065	.4221	19
42	.70339	.71080	.98958	.0105	.4069	.4217	18
43	.70360	.71059	.99016	.0099	.4073	.4212	17
44	.70381	.71039	.99073	.0093	.4077	.4208	16
45	.70401	.71018	.99131	1.0088	1.4081	1.4204	15
46	.70422	.70998	.99189	.0082	.4085	.4200	14
47	.70443	.70977	.99246	.0076	.4089	.4196	13
48	.70463	.70957	.99304	.0070	.4093	.4192	12
49	.70484	.70936	.99362	.0064	.4097	.4188	11
50	.70505	.70916	.99420	1.0058	1.4101	1.4183	10
51	.70525	.70895	.99478	.0052	.4105	.4179	9
52	.70546	.70875	.99536	.0047	.4109	.4175	8
53	.70566	.70854	.99593	.0041	.4113	.4171	7
54	.70587	.70834	.99651	.0035	.4117	.4167	6
55	.70608	.70813	.99709	1.0029	1.4122	1.4163	5
56	.70628	.70793	.99767	.0023	.4126	.4159	4
57	.70649	.70772	.99826	.0017	.4130	.4154	3
58	.70669	.70752	.99884	.0012	.4134	.4150	2
59	.70690	.70731	.99942	.0006	.4138	.4146	1
60	.70711	.70711	1.0000	1.0000	1.4142	1.4142	0
M	**Cosine**	**Sine**	**Cotan.**	**Tan.**	**Cosec.**	**Secant**	**M**

45°

PIPE AND WATER WEIGHT PER LINE FOOT

Nom. Pipe Size	STD. WT. OF:		XS WEIGHT OF:	
	PIPE	WATER	PIPE	WATER
½	.851	.132	1.088	.101
¾	1.131	.231	1.474	.187
1	1.679	.374	2.172	.311
1¼	2.273	.648	2.997	.555
1½	2.718	.882	3.632	.765
2	3.653	1.453	5.022	1.278
2½	5.794	2.073	7.662	1.835
3	7.58	3.20	10.25	2.86
3½	9.11	4.28	12.51	3.85
4	10.79	5.51	14.99	4.98
5	14.62	8.66	20.78	7.88
6	18.98	12.51	28.58	11.29
8	28.56	21.68	43.4	19.8
10	40.5	34.1	54.7	32.3
12	49.6	49.0	65.4	47.0
14	54.6	59.7	72.1	57.5
16	62.6	79.1	82.8	76.5
18	70.6	101.2	93.5	98.3
20	78.6	126.0	104.1	122.8
24	94.6	183.8	125.5	179.9
30	118.7	291.0	157.6	286.0

FEET HEAD OF WATER TO PSI

FEET HEAD	POUNDS PER SQUARE INCH	FEET HEAD	POUNDS PER SQUARE INCH
1	.43	100	43.31
2	.87	110	47.64
3	1.30	120	51.97
4	1.73	130	56.30
5	2.17	140	60.63
6	2.60	150	64.96
7	3.03	160	69.29
8	3.46	170	73.63
9	3.90	180	77.96
10	4.33	200	86.62
15	6.50	250	108.27
20	8.66	300	129.93
25	10.83	350	151.58
30	12.99	400	173.24
40	17.32	500	216.55
50	21.65	600	259.85
60	25.99	700	303.16
70	30.32	800	346.47
80	34.65	900	389.78
90	38.98	1000	433.00

NOTE: One foot of water at 62°F equals .433 pound pressure per sq. in. To find the pressure per sq. in. for any feet head not given in the table above, multiply the feet head by .433.

NOTES: